ENVIRONMENTAL SCIENCE, ENGINEERING AND TECHNOLOGY

AVIATION AND CLIMATE CHANGE

ENVIRONMENTAL SCIENCE, ENGINEERING AND TECHNOLOGY

Additional books in this series can be found on Nova's website at:

https://www.novapublishers.com/catalog/index.php?cPath=23_29&seriesp=Environmental%20Science%2C%20Engineering%20and%20Technology&sort=2a&page=1

CLIMATE CHANGE AND ITS CAUSES, EFFECTS AND PREDICTION

Additional books in this series can be found on Nova's website at:

https://www.novapublishers.com/catalog/index.php?cPath=23_29 &seriesp=Climate%20Change%20and%20its%20Causes%2C%20Effects%20and%20Prediction&sort=2a&page=1

ENVIRONMENTAL SCIENCE, ENGINEERING AND TECHNOLOGY

AVIATION AND CLIMATE CHANGE

GEORGE T. BLUMENTHAL
EDITOR

Nova Science Publishers, Inc.
New York

For permission to use material from this book please contact us:
Telephone 631-231-7269; Fax 631-231-8175
Web Site: http://www.novapublishers.com

LIBRARY OF CONGRESS CATALOGING-IN-PUBLICATION DATA

Aviation and climate change / Editor, George T. Blumenthal.
p. cm.
Includes index.
ISBN 978-1-60876-757-1 (hardcover)
1. Aircraft exhaust emissions--Environmental aspects--United States. 2. Aeronautics, Commercial--Environmental aspects--United States. 3. Greenhouse gases--Environmental aspects--United States. 4. Climatic changes--United States. I. Blumenthal, George T.
TD195.A27A95 2009
363.738'74--dc22

2010012179

Published by Nova Science Publishers, Inc. ✦ New York

CONTENTS

PREFACE

According to the Intergovernmental Panel on Climate Change (IPCC), aviation currently accounts for about 2 percent of human-generated global carbon dioxide emissions, the most significant greenhouse gas and about 3 percent of the potential warming effect of global emissions that can affect the earth's climate, including carbon dioxide. This book discusses the IPCC's medium-range estimate forecasts that by 2050 the global aviation industry, including aircraft emissions, will emit about 3 percent of global carbon dioxide emissions and about 5 percent of the potential warming effect of all global human-generated emissions.

Chapter 1 – Aircraft emit greenhouse gases and other emissions, contributing to increasing concentrations of such gases in the atmosphere. Many scientists and the Intergovernmental Panel on Climate Change (IPCC)—a United Nations organization that assesses scientific, technical, and economic information on climate change— believe these gases may negatively affect the earth's climate. Given forecasts of growth in aviation emissions, some governments are taking steps to reduce emissions. In response to a congressional request, GAO reviewed (1) estimates of aviation's current and future contribution to greenhouse gas and other emissions that may affect climate change; (2) existing and potential technological and operational improvements that can reduce aircraft emissions; and (3) policy options for governments to help address commercial aircraft emissions.

GAO conducted a literature review; interviewed representatives of government agencies, industry and environmental organizations, airlines, and manufacturers, and interviewed and surveyed 18 experts in economics and aviation on improvements for reducing emissions from aircraft.

GAO is not making recommendations. Relevant agencies provided technical comments which were incorporated as appropriate and EPA said emissions standards can have a positive benefit to cost ratio and be an important part of policy options to control emissions.

Chapter 2 – Aviation plays a key role in the economic prosperity and lifestyle Americans enjoy. Our economy benefits greatly from the ability to move people and products all over the globe - quickly and safely. Aviation contributes to our quality of life - allowing us to visit friends and relatives, to travel, to experience new places, to shrink the borders of the world. The statistics are impressive. In 1903 the year of the Wright brothers' first flight, earth's population was 1.6 billion[1]; today, over 1.6 billion people use the world's airlines[2]. The air transport industry provides 28 million direct, indirect, and induced jobs worldwide[3]. And aircraft carry about 40% of the value of all world trade[4], driving the "just in time" deliveries critical to productivity improvements.

Air transport links our world and is a key tenet of continued economic development and security for the U.S. However, aviation also has environmental impacts – primarily noise and atmospheric emissions. While aircraft noise issues are better known, less focus has been placed on emissions. This paper provides a brief overview of important issues regarding aviation emissions.

Chapter 3 - This chapter is edited and excerpted testimony by Richard L. Altman before the House Transportation Committee on May 6, 2008.

Chapter 4 – This chapter is edited and excerpted testimony by James K. Coyne before the House Transportation Committee on May 6, 2008.

Chapter 5 – This chapter is edited and excerpted testimony by Daniel K. Elwell before the House Transportation Committee on May 6, 2008.

Chapter 6 – This chapter is edited and excerpted written testimony by Dr. David W. Fahey before the House Transportation Committee on May 6, 2008.

Chapter 7 - This chapter is edited and excerpted written testimony by James K. May before the House Transportation Committee on May 6, 2008.

Chapter 8 – This chapter is edited and excerpted written testimony by Gerald L. Dillingham before the House Transportation Committee on May 6, 2008.

Chapter 9 - Aircraft are a significant source of greenhouse gases—compounds that trap the sun's heat, with effects on the Earth's climate. In the United States, aircraft of all kinds are estimated to emit between 2.4% and 3.4% of the nation's total greenhouse gas (GHG) emissions, depending on whether one counts international air travel. The impact of U.S. aviation on climate change is perhaps twice that size when other factors are considered. These include the contribution of aircraft emissions to ozone formation, the water vapor and soot that aircraft emit, and the high altitude location of the bulk of aircraft emissions. Worldwide, aviation is projected to be among the faster growing GHG sources.

If Congress and the new Administration decide to regulate aircraft GHG emissions, they face several choices. The Administration could use existing authority under Section 231 of the Clean Air Act, administered by the Environmental Protection Agency. EPA has already been petitioned to do so by several states, local governments, and environmental organizations. Congress could address aviation or aviation fuels legislatively, through cap-and-trade or carbon tax proposals.

Among the legislative options, the cap-and-trade approach (setting an economy-wide limit on GHG emissions and distributing tradable allowances to emitters) has received the most attention. In the 110th Congress, most cap-and-trade bills would have included aviation indirectly, through emission caps imposed upstream on their source of fuel—the petroleum refining sector. By capping emissions upstream of air carriers and eventually lowering the cap as much as 80%, bills such as these would have several effects: they would provide an incentive for refiners to produce lower carbon fuels; they would increase the price of fuels, and thus increase the demand for more fuel-efficient aircraft; and they might increase the cost of aviation services relative to other means of transport, giving airline passengers and shippers of freight incentives to substitute lower-cost, lower carbon alternatives.

Besides regulating emissions directly or through a cap-and-trade program or carbon tax, there are other tools available to policymakers that can lower aviation's GHG emissions. These include implementation of the Next Generation Air Traffic Control System (not expected to be complete until 2025, although some elements that could reduce aircraft emissions may be implemented sooner); research and development of more fuel-efficient aircraft and engines; and perhaps the development of lower carbon jet fuel.

This report provides background on aviation emissions and the factors affecting them; it discusses the tools available to control emissions, including existing authority under the Clean Air Act and proposed economy-wide cap-and-trade legislation; and it examines international regulatory developments that may affect U.S. commercial airlines. These include the European Union's Emissions Trading Scheme for greenhouse gases (EU-ETS), which is to include the aviation sector beginning in 2012, and discussions under the auspices of the International Civil Aviation Organization (ICAO).

In: Aviation and Climate Change
Editors: George T. Blumenthal

ISBN: 978-1-60876-757-1

Chapter 1

AVIATION AND CLIMATE CHANGE: AIRCRAFT EMISSIONS EXPECTED TO GROW, BUT TECHNOLOGICAL AND OPERATIONAL IMPROVEMENTS AND GOVERNMENT POLICIES CAN HELP CONTROL EMISSIONS

United States Government Accountability Office

WHY GAO DID THIS STUDY

Aircraft emit greenhouse gases and other emissions, contributing to increasing concentrations of such gases in the atmosphere. Many scientists and the Intergovernmental Panel on Climate Change (IPCC)—a United Nations organization that assesses scientific, technical, and economic information on climate change— believe these gases may negatively affect the earth's climate. Given forecasts of growth in aviation emissions, some governments are taking steps to reduce emissions. In response to a congressional request, GAO reviewed (1) estimates of aviation's current and future contribution to greenhouse gas and other emissions that may affect climate change; (2) existing and potential technological and operational improvements that can reduce aircraft emissions; and (3) policy options for governments to help address commercial aircraft emissions.

GAO conducted a literature review; interviewed representatives of government agencies, industry and environmental organizations, airlines, and manufacturers, and interviewed and surveyed 18 experts in economics and aviation on improvements for reducing emissions from aircraft.

GAO is not making recommendations. Relevant agencies provided technical comments which we incorporated as appropriate and EPA said emissions standards can have a positive benefit to cost ratio and be an important part of policy options to control emissions.

WHAT GAO FOUND

According to IPCC, aviation currently accounts for about 2 percent of human- generated global carbon dioxide emissions, the most significant greenhouse gas—and about 3 percent of the potential warming effect of global emissions that can affect the earth's climate, including carbon dioxide. IPCC's medium- range estimate forecasts that by 2050 the global aviation industry, including aircraft emissions, will emit about 3 percent of global carbon dioxide emissions and about 5 percent of the potential warming effect of all global human-generated emissions. Gross domestic product growth is the primary driver in IPCC's forecasts. IPCC also made other assumptions about future aircraft fuel efficiency, improvements in air traffic management, and airport and runway capacity. IPCC's 2050 forecasts for aviation's contribution to global emissions assumed that emissions from other sectors will continue to grow. If other sectors make progress in reducing emissions and aviation emissions continue to grow, aviation's relative contribution may be greater than IPCC estimated; on the other hand, if other sectors do not make progress, aviation's relative contribution may be smaller than estimated.

While airlines currently rely on a range of improvements, such as fuel-efficient engines, to reduce emissions, some of which may have limited potential to generate future reductions, experts we surveyed expect a number of additional technological, operational, and alternative fuel improvements to help reduce aircraft emissions in the future. However, according to experts we interviewed, some technologies, such as advanced airframes, have potential, but may be years away from being available, and developing and adopting them is likely to be costly. In addition, according to some experts we interviewed, incentives for industry to research and adopt low-emissions technologies will be dependent to some extent on the level and stability of fuel prices. Finally, given expected growth of commercial aviation as forecasted by IPCC, even if many of these improvements are adopted, it appears unlikely they would greatly reduce emissions by 2050.

A number of policy options to address aircraft emissions are available to governments and can be part of broader policies to address emissions from many sources including aircraft. Market-based measures can establish a price for emissions and provide incentives to airlines and consumers to reduce emissions. These measures can be preferable to other options because they would generally be more economically efficient. Such measures include a cap-and-trade program, in which government places a limit on emissions from regulated sources, provides them with allowances for emissions, and establishes a market for them to trade emissions allowances with one another, and a tax on emissions. Governments can establish emissions standards for aircraft or engines. In addition, government could increase government research and development to encourage development of low-emissions improvements.

ABBREVIATIONS

ACARE	Advisory Council for Aeronautics Research in Europe
ADS-B	Automatic Dependent Surveillance-Broadcast
ATA	Air Transport Association
CDA	Continuous Descent Arrival
CER	certified emissions reduction
CO_2	carbon dioxide
DG Environment	Directorate-General of the Environment
EEC	European Economic Community
EPA	Environmental Protection Agency
ERU	emission reduction unit
EU	European Union
EU	ETS European Union Emissions Trading Scheme
FAA	Federal Aviation Administration
GDP	gross domestic product
GIACC	Group on International Aviation and Climate Change
ICAO	International Civil Aviation Organization
IETA	International Emissions Trading Association

IPCC	Intergovernmental Panel on Climate Change
NASA	National Aeronautics and Space Administration
NextGen	Next Generation Air Transportation System
NOx	nitrogen oxides
RNAV	area navigation
RNP	Required Navigation Performance
SESAR	Single European Sky Air Traffic Management Research Program
UNFCCC	United Nations Framework Convention on Climate Change

June 8, 2009
Congressional Requesters

Many sources, including manufacturing, residential, and transportation sources, emit greenhouse gases that contribute to the accumulation of these gases in the earth's atmosphere. Greenhouse gases disperse and trap heat in the earth's atmosphere. This heat-trapping effect, known as the greenhouse effect, moderates atmospheric and surface temperatures, keeping the earth warm enough to support life. However, according to the Intergovernmental Panel on Climate Change (IPCC)—a United Nations organization that assesses scientific, technical, and economic information on the effects of climate change—global atmospheric concentrations of these greenhouse gases have increased markedly as a result of human activities over the past 200 years, contributing to a warming of the earth's climate. These trends, if unchecked, could have serious negative effects, such as rising sea levels and coastal flooding worldwide.

Aircraft emit a variety of greenhouse and other gases, including carbon dioxide—the most significant greenhouse gas emitted by aircraft—and nitrogen oxides, as well as other substances such as soot and water vapor that are believed to negatively affect the earth's climate. Airlines have a financial incentive to reduce carbon dioxide emissions, as those emissions are a direct result of fuel burn, which represents a large portion of their operating costs—about 30 percent for U.S. airlines in 2008. Some experts expect aviation to grow at a fast rate until 2021, when the Federal Aviation Administration (FAA) forecasts that U.S. domestic commercial aviation will serve over 1 billion passengers a year. While the current economic downturn could delay this growth somewhat, experts believe that growth in the aviation sector means greater productivity and mobility, but is also likely to increase emissions. To counteract expected increases in emissions, many governments and inter national organizations have set goals for future emissions reductions. For example, a number of developed countries have set a goal to reduce carbon dioxide emissions by 50 percent by 2050. In addition, the Kyoto Protocol, an international agreement to minimize the adverse effects of climate change, set binding targets for the reduction of greenhouse gases for 37 industrialized countries and the European Economic Community (EEC) to achieve during the 2008 through 2012 commitment period.[1] Although the United States is a signatory to the Kyoto Protocol, it is not bound by its terms or emissions target because it has not ratified the Protocol. The Protocol also requires industrialized nations and the EEC to pursue "limitations or reduction of emissions of greenhouses gases... from aviation... working through the International Civil Aviation Organization."[2] Finally, some governments have taken actions designed to control aviation emissions. For example, in 2003, the European Union (EU) established a cap-andtrade program known as the EU Emissions Trading Scheme (EU ETS) to control carbon dioxide emissions from various energy and industrial sectors. The EU ETS was first implemented in 2005 and was amended in 2008 to include aviation. Beginning in 2012, the ETS will include all covered flights into or out of an EU airport.[3]

You asked us to provide information on aviation emissions of greenhouse gases and other emissions that may affect climate change. To do so, we identified (1) aviation's current and estimated future contribution to the emissions of greenhouse gases and other emissions that may affect climate change, (2) existing and potential future technological and operational improvements that the commercial aviation industry can use to reduce commercial aircraft emissions, and (3) policy options for the U.S. government and other governments to help reduce commercial aviation emissions and the potential costs and benefits of each option. You also asked that we describe the EU's plans to add the aviation industry to its existing ETS and the potential legal implications of doing so. (See app. I for this description.) To address these objectives, we reviewed studies on the impact of aviation on climate change. We also collaborated with the National Academy of Sciences to identify and recruit experts with experience in climate change and the aviation industry. We interviewed 18 such experts (see app. II for a list of the experts). After these interviews, we asked the experts to complete a survey in which they assessed a list of options to reduce emissions on a variety of predetermined factors, such as potential for emissions reductions and costs (see app. III for complete results). In addition, we spoke with government, airline, and interest group officials in the United States, the EU, and the United

Kingdom, focusing on commercial aviation. (See app. IV for a more detailed description of our scope and methodology.) We conducted our work from March 2008 through June 2009 in accordance with generally accepted government auditing standards. Those standards require that we plan and perform the audit to obtain sufficient, appropriate evidence to provide a reasonable basis for our findings and conclusions based on our audit objectives. We believe that the evidence obtained provides a reasonable basis for our findings and conclusions based on our audit objectives.

BACKGROUND

Emissions from a variety of human-generated sources, including commercial aircraft, trap heat in the atmosphere and contribute to climate change. During flight operations, aircraft emit a number of greenhouse gas and other emissions, including carbon dioxide, nitrogen oxides (NOx), soot, and water vapor.

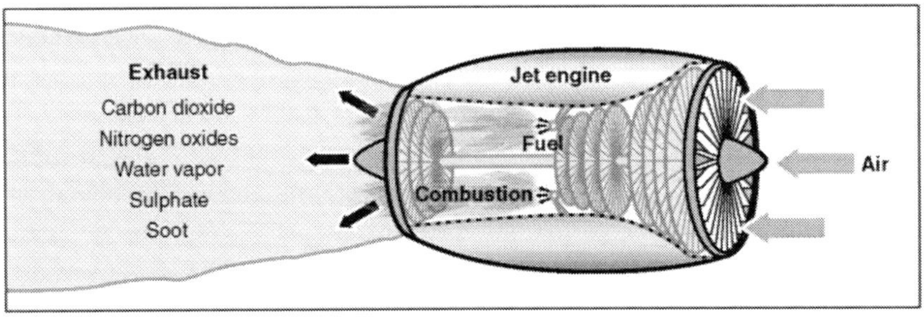

Source: GAO.

Figure 1. Selected Greenhouse Gas and Other Emissions from Aircraft at Cruising Altitude

Figure 1 shows the primary emissions from commercial aircraft. Carbon dioxide emissions from aircraft are a direct result of fuel burn. For every gallon of jet fuel burned, about 21 pounds of carbon dioxide are emitted. Reducing the amount of fuel burned, therefore, also reduces the amount of carbon dioxide emitted. Water vapor emissions and certain atmospheric temperature and humidity conditions can lead to the formation of contrails, a cloudlike trail of condensed water vapor, and can induce the creation of cirrus clouds. Both contrails and cirrus clouds are believed to have a warming effect on the earth's atmosphere. Aircraft also emit other pollutants that affect local air quality. Finally, airport operations are sources of greenhouse gas and other emissions, which we are not examining in this report.Historically, the commercial aviation industry has grown substantially in the United States and worldwide and is a contributor to economic growth. Between 1981 and 2008, passenger traffic increased 226 percent in the United States on a revenue passenger mile basis and 257 percent globally on a revenue passenger kilometer basis.[4] According to the FAA, in 2006 the civil aviation industry in the United States directly and indirectly contributed 11 million jobs and 5.6 percent of total gross domestic product (GDP) to the U.S. economy. Globally, the International Air Transport Association estimated that in 2007 the aviation industry had a global economic impact of over $3.5 trillion, equivalent to about 7.5 percent of worldwide GDP. Recently, however, the airline industry has experienced declining traffic and financial losses as the result of the current recession.

The fuel efficiency of commercial jet aircraft has improved over time. According to IPCC, aircraft today are about 70 percent more fuel efficient on a per passenger kilometer basis than they were 40 years ago because of improvements in engines and airframe design.[5] The cost of jet fuel is a large cost for airlines. In the 2008, when global fuel prices were high, jet fuel accounted for about 30 percent of U.S. airlines' total operating expenses, compared with 23 percent during 2007.

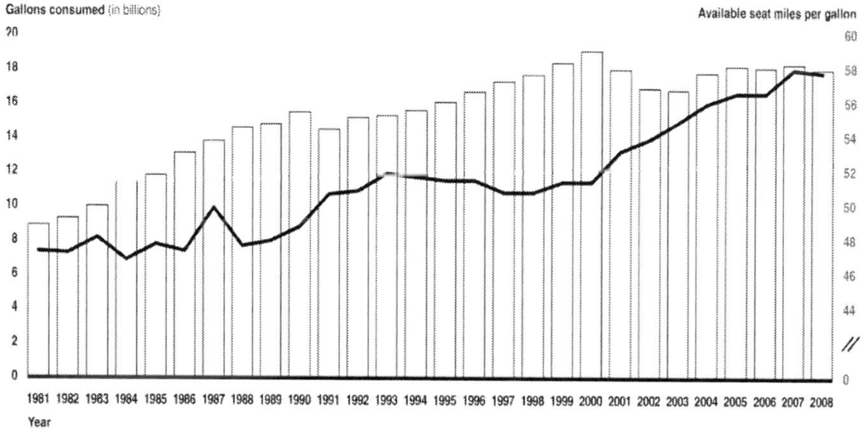

Source: U.S. Department of Transportation.

Figure 2. Total Fuel Consumption and Fuel Efficiency of U.S. Airlines

Fuel efficiency (measured by available seat-miles per gallon consumed) for U.S. carriers increased about 17 percent between 1990 and 2008, as shown in figure 2. Internationally, according to the International Air Transport Association, fuel efficiency (measured by revenue passenger kilometers) improved 16.5 percent between 2001 and 2007. According to FAA, between 2000 and early 2008 U.S. airlines reduced fuel burn and emissions while transporting more passengers and cargo.

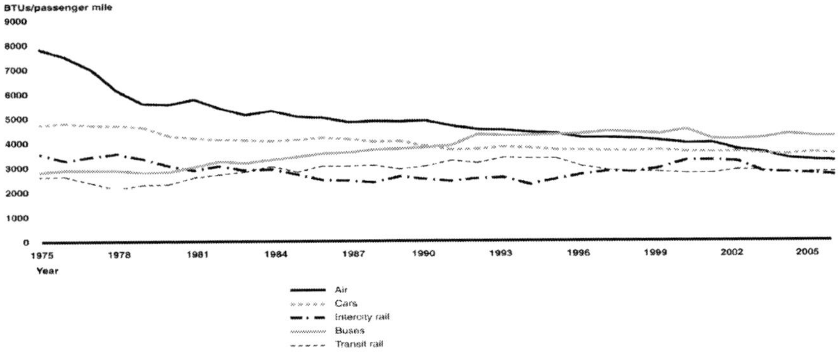

Source: Department of Energy.

Figure 3. Energy use per Passenger-mile, by Mode of Transportation

In addition, commercial aviation has become less energy intensive over time—that is, to transport a single passenger a single mile uses less energy than it previously did, measured in British thermal units. See figure 3 showing energy intensity over time of aviation and other modes of transportation.

However, despite these efficiency improvements, overall fuel burn and emissions of U.S. airlines are expected to grow in the future. FAA forecasts that between 2008 and 2025 fuel consumption of U.S.-based airlines will increase an average of 1.6 percent per year while revenue passenger miles will increase an average of 3.1 percent per year over the same period. As seen in figure 4, FAA forecasts that between 2008 and 2025 fuel consumption of U.S.-based airlines will increase an average of 1.6 percent per year.

To develop a better understanding of the effects of human-induced climate change and identify options for adaptation[6] and mitigation,[7] two United Nations organizations established IPCC in 1988 to assess scientific, technical, and socio-economic information on the effects of climate change. IPCC releases and periodically updates estimates of future greenhouse gas emissions from human activities under different economic development scenarios. In 1999, IPCC released its report, *Aviation and the Global Atmosphere, conducted at the request of the International Civil Aviation Organization* (ICAO)—a United Nations organization that aims to promote the establishment of international civilian aviation standards and recommended practices and procedures. In 2007, IPCC released an update on emissions from transportation and other sectors called the *Fourth Assessment Report of the Intergovernmental Panel on Climate Change*. These reports were developed with input from over 300 experts worldwide and are internationally accepted and used for policy-making.

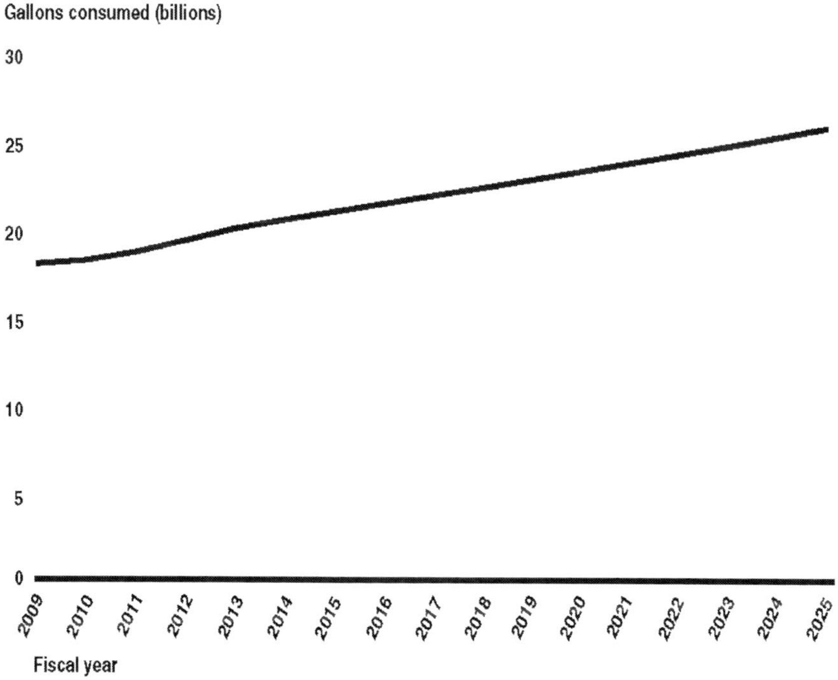

Source: FAA Aerospace Forecast, FY 2009-2005.

Figure 4. Forecasted Fuel Consumption by U.S. Airlines

A variety of federal agencies have roles in addressing aviation emissions. In 2004, FAA and other organizations including the National Aeronautics and Space Administration (NASA) released a report, *Aviation and the Environment: A National Vision Statement, Framework for Goals and Recommended Actions*, through the collaborative PARTNER program,[8] stating a general goal to reduce overall levels of emissions from commercial aviation and proposing actions to deal with aviation emissions. FAA also is involved in a number of emissions-reduction initiatives—including work on low-emissions technologies and low-carbon alternative fuels; the implementation of a new air traffic management system, the Next Generation Air Trans portation System (NextGen);[9] and climate research to better understand the impact of emissions from aviation. NASA has been involved in research that has led to the development of technologies that reduce aircraft emissions. Currently, NASA's Subsonic Fixed-Wing project, part of its Fundamental Aeronautics program, aims to help develop technologies to reduce fuel burn, noise, and emissions in the future. Both FAA and NASA are involved in the Aviation Climate Change Research Initiative, whose goals include improving the scientific understanding of aviation's impact on climate change. Also, as mandated under Title II of the Clean Air Act, the Environmental Protection Agency (EPA) promulgates certain emissions standards for aircraft and aircraft engines[10] and has adopted emission standards matching those for aircraft set by ICAO.[11] While neither ICAO nor EPA has established standards for aircraft engine emissions of carbon dioxide, ICAO is currently discussing proposals for carbon dioxide emissions standards and considering a global goal for fuel efficiency.[12] In addition, in 2007 a coalition of environmental interest groups filed a petition with EPA asking the agency, pursuant to the Clean Air Act, to make a finding that "greenhouse gas emissions from aircraft engines may be reasonably anticipated to endanger the public health and welfare" and, after making this endangerment finding, promulgate regulations for greenhouse gas emissions from aircraft engines.[13]

International concerns about the contribution of human activities to global climate change have led to several efforts to reduce their impact. In 1992, the United Nations Framework Convention on Climate Change (UNFCCC)—a multilateral treaty whose objective is to stabilize greenhouse gas concentrations in the atmosphere at a level that would prevent dangerous human interference with the climate system—was signed.[14] By 1995, the parties to the UNFCCC, including the United States, realized that progress toward this goal was not sufficient. In December 1997, the parties reconvened in Kyoto, Japan, to adopt binding measures to reduce greenhouse gas emissions. Under the resulting Kyoto Protocol, which the United States has not ratified, industrialized nations committed to reduce or limit their emissions of carbon dioxide and other greenhouse gases during the 2008 through 2012 commitment period.[15] The Protocol directed the industrialized nations to work through ICAO to reduce or limit emissions from aviation, but international aviation emissions are not explicitly included in Kyoto's targets. In 2004, ICAO endorsed the further development of an open emissions trading system for international aviation, and in 2007 called for mutual agreement between contracting states before implementation of an emissions trading scheme. In part to meet its Kyoto Protocol requirements, the EU implemented its ETS in 2005, which sets a cap on carbon dioxide emissions and allows regulated entities to buy and sell emissions allowances with one another. In 2008, the European Parliament and the Council of the European Union passed a directive, or law, to include aviation in the ETS. Under the directive, beginning in 2012 a cap will be placed on total carbon dioxide emissions from all covered flights by aircraft operators into or out of an EU airport.[16] Many stakeholders and countries have stated objections to the EU's plans and legal challenges are possible. (See app. I for a discussion of the ETS's inclusion of aviation.) In December 2009, the parties to the UNFCCC will convene in Copenhagen, Denmark, to discuss and negotiate a post-Kyoto framework for addressing global climate change.

AVIATION EMISSIONS REPRESENT A SMALL BUT GROWING SHARE OF ALL EMISSIONS

IPCC estimates that aviation emissions currently account for about 2 percent of global human-generated carbon dioxide emissions and about 3 percent of the radiative forcing[17] of all global human-generated emissions (including carbon dioxide) that contribute to climate change. On the basis of available data and assumptions about future conditions, IPCC forecasted emissions to 2015 and forecasted three scenarios—low, medium, and high—for growth in global aviation carbon dioxide emissions from 2015 to 2050. These scenarios are driven primarily by assumption about economic growth—the factor most closely linked historically to the aviation industry's growth—but they also reflect other aviation-related assumptions. Because IPCC's forecasts depend in large part on assumptions, they, like all forecasts, are inherently uncertain. Nevertheless, as previously noted, IPCC's work reflects the input of over 300 leading and contributing authors and experts worldwide and is internationally accepted and used for policy making.[18]

Aviation Contributes about 2 Percent of Global Carbon Dioxide Emissions

According to IPCC, global aviation contributes about 2 percent of the global carbon dioxide emissions caused by human activities.[19] This 2 percent estimate includes emissions from all global aviation, including both commercial and military. Global commercial aviation, including cargo, accounted for over 80 percent of this estimate.[20] In the United States, domestic aviation contributes about 3 percent of total carbon dioxide emissions, according to EPA data.[21]

Many industry sectors, such as the electricity-generating and manu facturing sectors, contribute to global carbon dioxide emissions, as do residential and commercial buildings that use fuel and power. The transportation sector also contributes substantially to global carbon dioxide emissions. Specifically, it accounts for about 20 percent of total global carbon dioxide emissions.[22] Road transportation accounts for the largest share of carbon dioxide emissions—74 percent—from the transportation sector; aviation accounts for about 13 percent of carbon dioxide emissions from all transportation sources; and other transportation sources, such as rail, account for the remaining 13 percent. Figure 5 shows the relative contributions of industry, transportation, and all other sources to global carbon dioxide emissions and breaks down transportation's share to illustrate the relative contributions of road traffic, aviation, and other transportation sources.

Aviation Contributes about 3 Percent of All Human- Generated Emissions

When other aviation emissions—such as nitrogen oxides, sulfate aerosols, and water vapor—are combined with carbon dioxide, aviation's estimated share of global emissions increases from 2 percent to 3 percent, according to IPCC. However, the impact of these other emissions on climate change is less well understood than the impact of carbon dioxide, making IPCC's combined estimate more uncertain than its estimate for carbon dioxide alone.

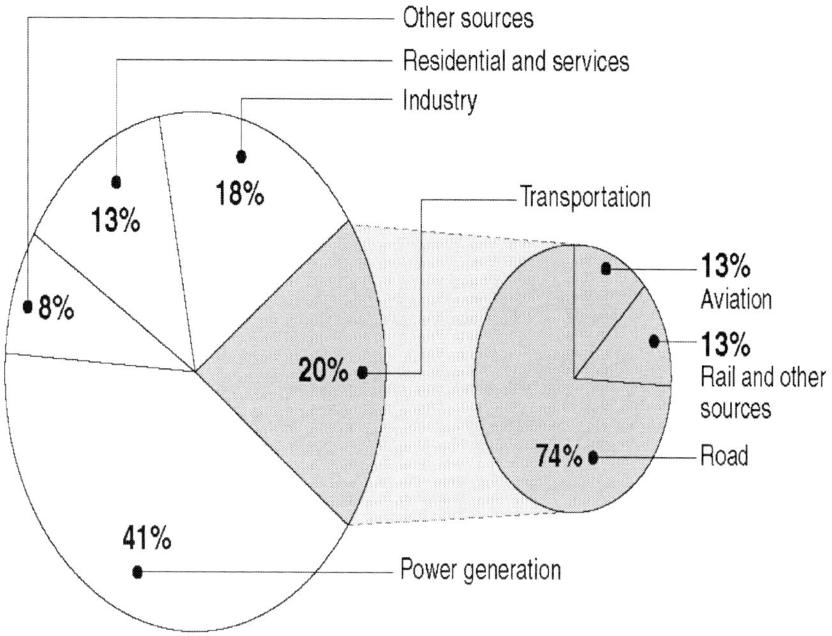

Source: GAO presentation of IPCC reported study.

Figure 5. Global Transportation's and Global Aviation's Contributions to Carbon Dioxide Emissions, 2004

Aviation emissions may contribute directly or indirectly to climate change. Although most aviation emissions have a warming effect, sulfate aerosols and a chemical reaction involving methane have a cooling effect. The warming effect is termed "positive radiative forcing" and the cooling effect "negative radiative forcing." Aviation emissions also may contribute to the formation of cirrus clouds, which can cause atmospheric warming, but the scientific community does not yet understand this process well enough to quantify the warming effect of aviation-induced cirrus clouds. Table 1 describes the direct or indirect effects of aviation emissions on climate change.

According to IPCC, when the positive radiative forcing effects of carbon dioxide and the positive and negative radiative forcing effects of other aviation emissions are combined, global aviation contributes about 3 percent[23] of human-generated positive radiative forcing. When the radiative forcing effects of the various aviation emissions are considered , carbon dioxide, nitrogen oxides, and contrails have the greatest potential to contribute to climate change.

Table 1. Types of Aviation Emissions and Their Effects at Cruising Altitude

Direct greenhouse gases and emissions	
Carbon dioxide	Carbon dioxide has a warming effect on the climate and remains in the atmosphere for hundreds of years. Carbon dioxide emissions from aviation have the same effect as those from other industry sectors because the carbon dioxide emitted from aircraft remains in the atmosphere long enough to be well mixed with the carbon dioxide emitted from ground-based sources.
Water vapor	Water vapor has a warming effect on the climate and is generated from the hydrogen contained in aviation fuel. It remains for only a short period in the troposphere (the lowest portion of the earth's atmosphere), where most emissions from aviation occur. The quantity of water vapor emitted by aviation is small compared with the quantities emitted from natural atmospheric sources.
Soot particles	Soot particles are produced during combustion and have a small warming (positive radiative forcing[a]) effect on the climate as they absorb incoming sunlight and heat the atmosphere.
Indirect greenhouse gases and emissions	
Ozone (from NOx)	Nitrogen emissions do not contribute to global warming directly, but the ozone generated from increased nitrogen oxides (NOx) is a greenhouse gas that produces a warming effect on the climate. The effect is higher at cruising altitude than on the ground because of longer lifetimes and greater radiative forcing in the higher levels of the atmosphere.
Methane (NOx related)	Very little or no methane is emitted by aircraft, but NOx emissions initiate a destruction ofmethane molecules, which generates an overall cooling effect on the climate. Overall, the warming effect of NOx emissions due to ozone formation is estimated to be higher than the cooling that results from methane destruction.
Sulfate aerosols	Sulfate aerosols, which arise from sulfur in jet fuel, scatter incoming sunlight back to the atmosphere and have a relatively small cooling effect on the climate.
Cloud formation	
Contrails	Contrails are formed through emissions of water and particles under certain atmospheric conditions such as temperature and humidity. They mainly consist of water already contained in the atmosphere, and aircraft operations only trigger their formation in these areas. Contrails cool the climate through increased

Table 1. (Continued)

Cloud formation	
	reflection of solar radiation, but also trap heat on the earth, which contributes to global warming. Overall, contrails have a warming effect, although uncertainties about the magnitude of this effect remain.
Cirrus clouds	Cirrus cloud formation might be augmented through aviation-induced contrails and cloud seeding from emission particles. Cirrus clouds have a warming effect, but exact quantifications are not yet possible. Consequently, IPCC did not include the impact of cirrus cloud formation in its estimates of aviation's contribution to human-generated emissions.[b]

Source: IPCC, the Aviation Climate Change Research Initiative, and the European Topic Center on Air and Climate Change.

[a] According to the IPCC, radiative forcing has advantages over global warming potential (GWP) for estimating aviation's impact from short-lived aviation emissions and aerosols and contrails. GWP is used to measure the heat-absorbing ability of long-lived gases using carbon dioxide as a reference. Each measure has limits and several other impact measures including a modified GWP are under consideration to address some of the limitations.

[b] University and government organizations, including NASA, are conducting research on them to better understand their effects.

The level of scientific understanding about the impact of particular aviation emissions on radiative forcing varies, making estimates of their impact on climate change uncertain to varying degrees. A recent report that described levels of scientific understanding of aviation emissions found that the levels for carbon dioxide were high; the levels for nitrogen oxides, water vapor, sulfates, and soot were medium; and the levels for contrails and aviation-induced cirrus clouds were low.[24] Aviation's contribution to total emissions, estimated at 3 percent, could be as low as 2 percent or as high as 8 percent, according to IPCC. Figure 6 shows IPCC's estimate of the relative positive radiative forcing effects of each type of aviation emission for the year 2000. The overall radiative forcing from aviation emissions is estimated to be approximately two times that of carbon dioxide alone.[25]

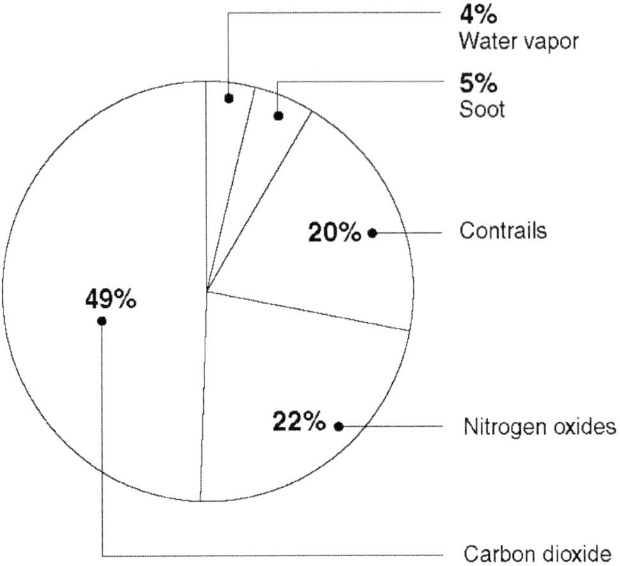

Source: GAO presentation of IPCC reported study.

Note: This figure portrays the relative contributions of emissions that have a net positive radiative forcing effect. The effect of sulfate emissions, which have a negative radiative forcing, or cooling, effect, is not included. The result reported for nitrogen oxides is the net warming effect calculated by subtracting the cooling effect of methane from the warming effect of ozone. The aviation radiative forcing impact estimates described by IPCC in 2007 are reported as (positive) warming and (negative) cooling in milliwatts per meter squared: carbon dioxide (+ 25.3), ozone production from nitrogen oxides (+21.9), reduction of atmospheric methane as a result of nitrogen oxides (-10.4), water vapor (+2.0), sulfate particles (-3.5), soot particles (+2.5), and contrails (+10.0). For cirrus clouds, IPCC had no best estimate due to uncertainty, but a possible range of +10 to +80 was reported. The relative contributions of emissions in this chart are an approximation because of the uncertainty surrounding the non-carbon dioxide forcing estimates.

IPCC, 2007: Climate Change 2007: The Physical Science Basis. Contribution of Working Group I to the Fourth Assessment Report of the Intergovernmental Panel on Climate Change [Solomon, S., D. Qin, M. Manning, Z. Chen, M. Marquis, K.B. Averyt, M. Tignor and H.L. Miller (eds.)]. Cambridge University Press, Cambridge, United Kingdom and New York, NY, USA.

Figure 6. Estimated Relative Contribution of Aviation Emissions to Positive Radiative Forcing

Global Aviation Emissions Are Expected to Grow but Forecasts Vary, Primarily Reflecting Different Economic Growth Assumptions

IPCC generated three scenarios that forecasted the growth of global aviation carbon dioxide emissions from the near-term (2015) to the longterm (2050) and described these scenarios in its 1999 report. These forecasts are generated by models that incorporate assumptions about future conditions, the most important of which are assumptions about global economic growth and related increases in air traffic. Other assumptions include improvements in aircraft fuel efficiency and air traffic management and increases in airport and runway capacity. Because the forecasts are based on assumptions, they are inherently uncertain.

Forecasts of global economic growth and air traffic primarily drive IPCC's emissions estimates

Historically, global economic growth has served as a reliable indicator of air traffic levels. Aviation traffic has increased during periods of economic growth and slowed or decreased during economic slowdowns. As figure 7 shows, U.S and global passenger traffic (including the U.S.) generally trended upward from 1978 through 2008, but leveled off or declined during economic recessions in the United States.

Forecast models described in IPCC's report incorporate historical trends and the relationship between economic growth and air traffic to produce scenarios of global aviation's potential future carbon dioxide emissions. IPCC used a NASA emissions forecast for carbon dioxide emissions until 2015. IPCC used an ICAO emissions forecasting model to forecast emissions from 2015 to 2050 using three different assumptions for global economic growth—low (2.0 percent), medium (2.9 percent), and high (3.5 percent). As a result, IPCC produced three different potential scenarios for future air traffic and emissions.[26] The 2050 scenarios include a 40 percent to 50 percent increase in fuel efficiency by 2050 from improvements in aircraft engines and airframe technology and from deployment of an advanced air traffic management system (these are discussed in more detail below). Figure 8 shows IPCC's low-, mid-, and high-range scenarios for carbon dioxide emissions for 2015, 2025, and 2050 as a ratio over 1990 emissions. IPCC used the medium economic growth rate scenario to estimate aviation's contribution to overall emissions in 2050.

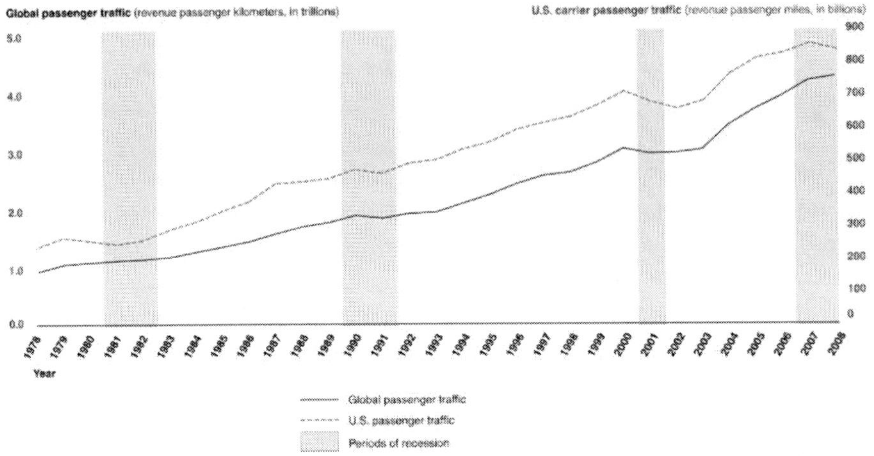

Sources: U.S. Department of Transportation data and ICAO data from the Air Transport Association.

Figure 7. Changes in Global and U.S. Aviation Passenger Traffic, 1978 through 2008

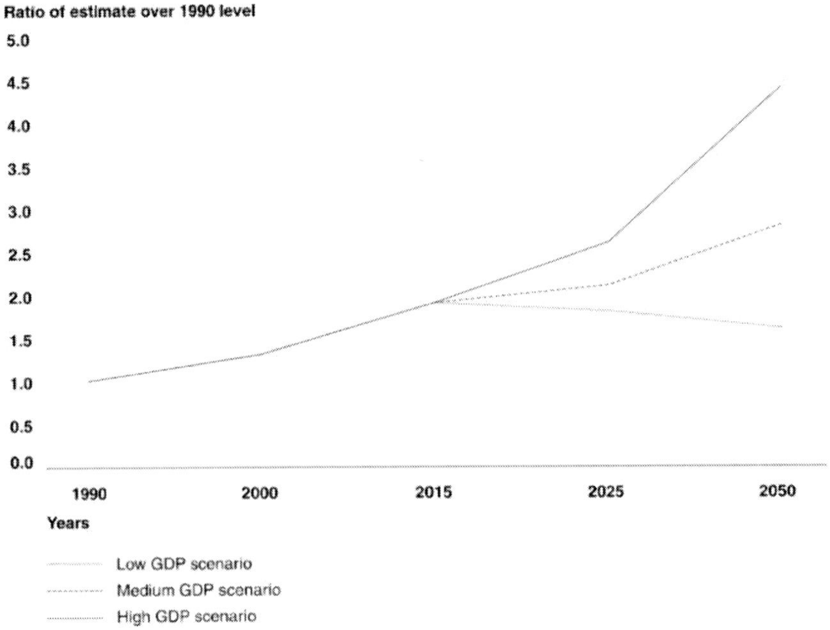

Source: IPCC.

Note: Includes advanced air traffic management and 40 percent to 50 percent fuel efficiency gain by 2050.

Figure 8. IPCC's Scenarios for Global Aviation Carbon Dioxide Emissions

IPCC compared aviation and overall emissions for the future and found that global aviation carbon dioxide emissions could increase at a greater rate than carbon dioxide emissions from all other sources of fossil fuel combustion. For example, for the medium GDP growth rate scenario, IPCC assumed a 2.9 percent annual average increase in global GDP, which translated into almost a tripling (a 2.8 times increase) of aviation's global carbon dioxide emissions from 1990 to 2050. For the same medium GDP growth scenario, IPCC also estimated a 2.2 times increase of carbon dioxide emissions from all other sources of fossil fuel consumption worldwide during this period. Over all, using the midrange scenario for global carbon dioxide emissions and projections for emissions from other sources, IPCC estimated that in 2050, carbon dioxide emissions from aviation could be about 3 percent of global carbon dioxide emissions, up from 2 percent. IPCC further estimated that, when other aviation emissions were combined with carbon dioxide emissions, aviation would account for about 5 percent of global human-generated positive radiative forcing, up from 3 percent.[27] IPCC concluded that the aviation traffic estimates for the low-range scenario, though plausible, were less likely given aviation traffic trends at the time the report was published in 1999. IPCC's 2007 *Fourth Assessment Report* included two additional forecasts of global aviation carbon dioxide emissions for 2050 developed through other studies.[28] Both of these studies forecasted mid- and high-range aviation carbon dioxide emissions for 2050 that were within roughly the same range as the 1999 IPCC report's forecasts.[29] For example, one study using average GDP growth assumptions that were similar to IPCC's showed mid- and high-range estimates that were close to IPCC's estimates.

In 2005, FAA forecasted a 60 percent growth in aviation carbon dioxide and nitrogen oxide emissions from 2001 to 2025. However, FAA officials recently noted that this estimate did not take into account anticipated aircraft fleet replacements, advances in aircraft and engine technology, and improvements to the air transportation system, nor did it reflect the recent declines in air traffic due to the current recession. After taking these factors into account, FAA reduced its estimate in half and now estimates about a 30 percent increase in U.S. aviation emissions from 2001 to 2025.[30] To account for some uncertainties in FAA's emissions forecasting, FAA officials said they are working on creating future scenarios for the U.S. aviation sector to assess the influence of a range of technology and market assumptions on future emissions levels.

Other forecasts show continued long-term growth, but emissions could fall below estimated levels during the current economic downturn

While recent aviation forecasts are generally consistent with IPCC's expectation for long-term global economic growth, the current economic slowdown has led to downward revisions in growth forecasts. For example, in 2008, Boeing's annual forecast for the aviation market projected a 3.2 percent annual global GDP growth rate from 2007 to 2027. However, this estimate was made before the onset of negative global economic growth in 2009 and could be revised downward in Boeing's 2009 forecast. According to FAA's March 2009 Aerospace Forecast, global GDP, which averaged 3 percent annual growth from 2000 to 2008, will be 0.8 percent from 2008 to 2010 before recovering to an estimated average annual growth rate of 3.4 percent from 2010 to 2020.[31] The International Air Transport Association has predicted that global air traffic will decrease by 3 percent in 2009 with the economic downturn. Moreover, according to the association, even if air traffic growth resumes in 2010, passenger air traffic levels will be 12 percent lower in the first few years after the slowdown and 9 percent lower in 2016 than the association forecasted in late 2007. To the extent that air traffic declines, emissions also will decline.

Assumptions about other factors could affect IPCC's forecasts

In developing its forecasts, IPCC made assumptions about factors other than economic growth that also affected its forecast results, as IPCC itself, experts we interviewed, and FAA have noted:

- IPCC assumed that advances in aircraft technology and the introduction of new aircraft would increas e fuel efficiency by 40 percent to 50 percent from 1997 through 2050.[32]
- IPCC assumed that an ideal air traffic management system would be in place worldwide by 2050,[33] reducing congestion and delays.
- IPCC assumed that airport and runway capacity would be sufficient to accommodate future air traffic levels.

However, if IPCC's assumptions about improvements in fuel efficiency and air traffic management are not realized, aircraft could produce higher emissions levels than IPCC estimated and IPCC's estimates would be understated. Conversely, if airports and runways have less capacity than IPCC assumed, then air traffic levels could be lower and, according to IPCC and some experts, IPCC's forecast could overstate future aviation emissions.

Finally, IPCC pointed out that its estimate that aviation will contribute 5 percent of positive radiative forcing in 2050 does not include the potential impact of aviation-induced cirrus clouds, which could be substantial.[34]

Because IPCC's forecasts are based on assumptions about future conditions and scientific understanding of the radiative forcing effects of certain aviation emissions is limited, IPCC's forecasts are themselves uncertain. According to FAA officials, given the numerous assumptions and inherent uncertainties involved in forecasting aviation emissions levels out to the year 2050, along with the significant shocks and structural changes the aviation community has experienced over the last few years, IPCC's projections are highly uncertain, even for the midrange scenario. If emissions from aviation and all other sectors continue to grow at about the same relative rate, aviation's contribution as a portion of overall emissions will not change significantly. However, if significant reductions are made in overall emissions from other sources and aviation emission levels continue to grow, aviation's contribution could grow.

Experts Believe Future Technological and Operational Improvements Are Likely to Help Reduce Emissions from Commercial Aircraft, but Likely Not by Enough to Fully Offset Estimated Market Growth

According to experts we interviewed, a number of different technological and operational improvements related to engines, aircraft design, operations, next-generation air traffic management, and fuel sources are either available now or are anticipated in the future to help reduce carbon dioxide emissions from aircraft. We interviewed and surveyed 18 experts in the fields of aviation and climate change and asked them to assess a number of improvements to reduce emissions using a variety of factors, such as potential costs and benefits, and then used the results to inform the following discussion. (Complete survey results can be found in app. III.) The development and adoption of low-emissions technologies is likely to be dependent upon fuel prices or any government policies that price aircraft emissions. Higher fuel prices or prices on emissions—for example through government policies such as an emissions tax—would make the costs of low-emissions technologies relatively cheaper and are likely to encourage their development. In addition, while fuel efficiency and emissions reductions may be important to airlines, so are a number of other factors, including safety, performance, local air quality, and noise levels, and trade-offs may exist between these factors.

Experts Believe That Although Many Technologies Are Expected to Help Reduce Emissions Growth in the Future, They Involve Trade-offs

Aircraft Engine Improvements

Improvements to aircraft engines have played a primary role in increasing fuel efficiency and reducing engine emission rates; experts we interviewed expect them to do so in the future—one study estimates that 57 percent of improvements in aircraft energy intensity between 1959 and 1995 were due to improvements in engine efficiency.[35] Such improvements have resulted from increasing engine pressure and temperatures (which increases their efficiency and decreases fuel usage) and improving the "bypass ratio," a measure of airflow through the engine.[36] However, according to experts we surveyed, further advances in these technologies may face high development costs (see table 2), and some may not be available for commercial use any time soon because engineers still face challenges in improving engine technology.

Some technologies may be available sooner than others, but all present a range of challenges and tradeoffs:

- One latest-generation aircraft engine, the geared turbofan engine, is likely to be available for use in certain aircraft in the next few years; promises to reduce emissions according to its manufacturer, Pratt & Whitney; and may face few challenges to widespread adoption.[37] According to Pratt & Whitney, this engine design is estimated to reduce fuel burn and emissions by 12 percent, compared with similar engines now widely used, in part due to an increase in the engine's bypass ratio. The geared turbofan engine is the result of research conducted by NASA and Pratt & Whitney.[38]
- Another engine technology, which could be introduced in the next 5 to 15 years, is the "open rotor" engine. It may deliver even greater emissions reductions but may face consumer-related challenges. The open rotor engine holds the engine fan blades on the outside of the engine case, thereby increasing the air flow around the engine, the effective bypass ratio, and the efficiency of the engine's propulsion.

However, this engine may be noisy and its large, visible engine blades could raise consumer concerns according to experts we surveyed. Research in the United States is currently a joint effort of NASA and General Electric. Rolls-Royce is also pursuing this technology.

- In the longer term, despite some engineering challenges, distributed propulsion technologies also hold promise for reducing aircraft emissions. Distributed propulsion systems would place many small engines throughout an aircraft instead of using a few large engines, as today's aircraft do. Experts we interviewed said that engineering challenges must be overcome with distributive propulsion, including determining the best and most efficient way to distribute power and store fuel. NASA is currently involved in distributed propulsion research.

Table 2. Selected Potential Aircraft Engine Improvements to Reduce Emissions

Improvement	Potential reduction in carbon dioxide emissions[a]	Potential research and development costs[a]	Estimated time frame for commercial use[b]	Potential forpublic acceptance[a]
Geared turbofan engine—more fuel efficient engine	Medium	Medium	Short-medium	High
Open rotor engine—engine fan blades not enclosed	High	High	Medium	Low-medium
Distributed propulsion systems—many small engines instead of few large ones	Medium	High	Long	High

Source: GAO survey of experts.

Note: Between 8 and 13 experts responded to each of our questions about technological improvements.

[a] We did not provide definitions for "low," "medium," or "high" in these cases.

[b] Short timeframe (<5 years); medium timeframe (5-15 years); long timeframe (>15 years).

Aircraft Improvements

Aircraft improvements also have played a role in reducing emissions rates in the past and experts we interviewed expected them to continue to do so. Through improvements in materials used to build aircraft and other improvements that increase aerodynamics and reduce drag, aircraft have become more fuel efficient over time. In the short term, improvements in aircraft materials, leading to decreased weight, and improvements in aerodynamics will help reduce fuel consumption and, thus, emissions rates. In the longer term, new aircraft designs, primarily a blended wing-body aircraft, hold potential for greater reductions in emissions rates. However, new aircraft concepts face engineering and consumer acceptance challenges and new technologies are likely to incur high development costs (see table 3).

Table 3. Selected Aircraft Improvements to Reduce Emissions

Improvement	Potential reduction in carbon dioxide emissions[a]	Potential research and development costs[a]	Estimated time frame for commercial use[b]	Potential for public acceptance[a]
Blended wing-body—Fuselage and wings as part of one airframe	High	High	Long	Low-medium
Lightweight composite airframes—Lightweight materials	Medium	Medium	Medium	High
Winglets—Wing attachments to reduce drag	Low	Low	Short[c]	High

Source: GAO survey of experts.

Note: Between 11 and 14 experts responded to each of our questions about technological improvements.

[a] We did not provide definitions for "low," "medium," or "high" in these cases.

[b] Short timeframe (<5 years); medium timeframe (5-15 years); long timeframe (>15 years).

[c] Winglets are already available and used by a number of airlines.

The following improvements to aircraft should help reduce aircraft fuel consumption and emissions in the long term, despite costs and challenges:

- The use of lightweight composite materials in aircraft construction has led to weight and fuel burn reductions in the past and is expected to continue to do so in the future. Over time, aircraft manufacturers have increasingly replaced more traditional materials such as aluminum with lighter-weight composite materials in airframe construction. For example, according to Boeing, 50 percent of the weight of the airframe of the Boeing 787, expected to be released in 2010, will be attributable to composite materials, compared with 12 percent composites in a currently available Boeing 777. According to Airbus, it first began using composite materials in airframe construction in 1985, and about 25 percent of the airframe weight of an A380 manufactured in 2008 was attributable to composites. By reducing the weight of the airframe, the use of composites reduces aircraft weight, fuel burn, and emissions rates.
- Retrofits such as winglets—wing extensions that reduce drag—can be made to aircraft to make them more aerodynamic but may have limited potential for future emissions reductions according to experts we surveyed. By improving airflow around wings, winglets reduce drag and improve fuel efficiency, thus reducing emissions by a modest amount. Boeing estimates that the use of winglets on a 737 reduces fuel burn by 3.5 percent to 4 percent on trips of over 1,000 nautical miles. Many new aircraft can be purchased with winglets, and existing aircraft also can be retrofitted with them. However winglets have already become very common on U.S. commercial airline aircraft and provide limited benefit for short-haul flights. According to experts we surveyed, there is low potential for future fuel consumption and emissions reductions from winglets.
- Redesigned aircraft, such as a blended wing-body aircraft—that is, an aircraft in which the body and wings are part of one airframe—hold greater potential for reducing emissions, according to experts we surveyed, though these face challenges as well. Several public and private organizations, including NASA and Boeing are conducting research on such aircraft. Many experts expect that blended wing-body aircraft will reduce emissions through improved aerodynamics and lighter weight. Estimates for potential emissions reductions include 33 percent compared with currently available aircraft

according to NASA. However, these new designs face challenges; notably, according to experts we interviewed, development costs are likely to be substantial, their radically different appearance may pose consumer acceptance issues, and they may require investments in modifying airports.[39]

Experts Also Expect Operational Improvements to Help Reduce Aircraft Emissions in the Future, but Reductions May Be Limited

Airlines have already taken a number of steps to improve fuel efficiency over time; however, the potential for future improvements from these measures may be limited. Airlines have increased their load factors (the percentage of seats occupied on flights), increasing the fuel efficiency of aircraft on a per-passenger basis. Load factors were about 80 percent for U.S. carriers in 2008, compared with about 65 percent in 1995. However, some experts we interviewed said the potential for additional future emissions reductions from increasing load factors may be small because they are already so high. Airlines also have removed many unnecessary items from aircraft and minimized supplies of certain necessary items, such as water, carried on board. As a result, according to some experts we interviewed, there may be little additional improvement in reducing emissions by reducing on-board weight. Airlines also have made other voluntary operational changes to reduce emissions, such as reducing speeds on certain routes, which reduces fuel use, and washing aircraft engines to make them cleaner and more efficient. Airlines also have retired less-fuel-efficient aircraft and replaced them with more-fuel-efficient models. For example, in 2008, American Airlines announced it was replacing more of its fuel-inefficient MD-80 aircraft with more efficient Boeing 737-800 aircraft. In addition, Continental Airlines, in 2008, replaced regional jets with turboprop planes on many routes. Still other improvements also are available for airlines to reduce emissions in the future, but the experts we interviewed ranked the potential for emissions reductions and consumer acceptance of these improvements as low (see Table 4).

Airlines could make other operational changes to reduce fuel burn and emissions but are unlikely to do so, because the potential for consumer acceptance of such changes is low according to experts we surveyed. For example, aircraft could fly in formation to improve airflow and reduce fuel burn. More specifically, rather than flying individually, several aircraft could fly in proximity to one another, reducing drag of aircraft and subsequently fuel use. However, aircraft would fly closer to one another than FAA's regulations currently allow and additional technological and aerodynamics research needs to be done. Another potential option, currently used for military purposes, is air-to-air refueling. Under this option, aircraft would be fueled in flight by tanker aircraft, reducing the amount and weight of fuel needed for the flight. However, DOT staff told us that air-to-air refueling may pose safety risks similar to those posed by formation flying. Some experts also have suggested that airlines make in- route on-ground fueling stops on long-haul flights, so they could reduce the amount of fuel they carry. However, more fueling stops could have negative effects on air quality at airports used for these stops as well as on air traffic operations.

Table 4. Selected Operational Improvements to Reduce Emissions

Improvement	Potential reduction in carbon dioxide emissions[a]	Potential research and development costs[a]	Estimated time frame for commercial use[b]	Potential for public acceptance[a]
Air-to-air refueling—Air tankers fueling aircraft in flight	Low	High	Long	Low
Engine washing—To improve engine performance	Low	Low	Short	High
Formation flying—Multiple aircraft flying close together to reduce drag	Low	Medium high	Medium-long	Low
Multi-stage long distance flights—Use of fueling stops on long-distance flights	Low-medium	Low	Short	Low

Source: GAO survey of experts.

Note: Between 9 and 12 survey respondents answered each of our questions about operational improvements to reduce emissions.

[a] We did not provide definitions for "low," "medium," or "high" in these cases.

[b] Short timeframe (<5 years); medium timeframe (5-15 years); long timeframe (>15 years).

Air Traffic Management Improvements through NextGen Will Incorporate Technological and Operational Improvements to Help Reduce Aircraft Emissions According to Experts

According to FAA, some of the air traffic management improvements that are part of NextGen—the planned air traffic management system designed to address the impacts of future traffic growth—can help reduce aircraft fuel consumption and emissions in the United States. Besides improving air traffic management, NextGen has environmental goals, which include accelerating the development of technologies that will lower emissions and noise. According to FAA, it is conducting a review to develop a set of NextGen goals, targets and metrics for climate change, as well as for noise and local air quality emissions. NextGen has the potential to reduce aircraft fuel burn by 2025, according to FAA, in part through technologies and procedures that reduce congestion and create more direct routing. Some procedures and technologies of NextGen have already been implemented and have already led to emissions reductions. Similarly, in Europe through the Single European Sky Air Traffic Management Research Program (SESAR), air traffic management technologies and procedures will be upgraded and individual national airspace systems will be merged into one, helping to reduce emissions per flight by 10 percent according to EUROCONTROL, the European Organization for the Safety of Air Navigation. However, some experts we met with said that because some of SESAR's technologies and procedures have already been implemented, future fuel savings might be lower. Table 5 provides information on selected components of NextGen that hold potential for reducing aircraft emissions.

NextGen has the potential to reduce fuel consumption and emissions through technologies and operational procedures:

- NextGen makes use of air traffic technologies to reduce emissions. For example, the Automatic Dependent Surveillance-Broadcast (ADS-B) satellite navigation system is designed to enable more precise control of aircraft during flight, approach, and descent, allowing for more direct routing and thus reducing fuel consumption and emissions. Also, Area Navigation (RNAV) will compute an aircraft's position and ground speed and provide meaningful information on the flight route to pilots, enabling them to save fuel through improved navigational capability. NextGen Network-Enabled Weather will provide real-time weather data across the national airspace system, helping reduce weather-related delays and allowing aircraft to best use weather conditions to improve efficiency.

- NextGen also relies on operational changes that have demonstrated the potential to reduce fuel consumption and emissions rates. Continuous Descent Arrivals (CDA) allow aircraft to remain at cruise altitudes longer as they approach destination airports, use lower power levels, and therefore produce lower emissions during landings. CDAs are already in place in a number of U.S. airports and according to FAA, the use of CDAs at Atlanta Hartsfield International Airport reduces carbon dioxide emissions by an average of about 1,300 pounds per flight. Required Navigation Performance (RNP) also permits an aircraft to descend on a more precise route, reducing its consumption of fuel and lowering its carbon dioxide emissions. According to FAA, over 500 RNAV and RNP procedures and routes have been implemented. Funding and other challenges, however, affect FAA's implementation of these various NextGen procedures and technologies.[40]

Table 5. Selected Air Traffic Management Improvements to Reduce Emissions

Improvement	Potential reduction in carbon dioxide emissions[a]	Potential research and development costs[a]	Estimated time frame for commercial use[b]	Potential for public acceptance[a]
Required navigation performance—More precise routes	Medium	Medium	Medium	High
Automatic Dependent Surveillance-Broadcast—Satellite navigation system	Medium	Medium	Short	High
Continuous Descent Arrival—More fuel efficient landings	Low-Medium	Medium	Short	High
NextGen Network-Enabled Weather—Advanced real-time weather data	Medium	Medium	Medium	High

Source: GAO survey of experts,

Note: 7 to 13 experts answered each of our survey questions about NextGen improvements

[a] We did not provide definitions for "low," "medium," or "high" in these cases.

[b] Short timeframe (<5 years); medium timeframe (5-15 years); long timeframe (>15 years)

Alternative Fuel Sources Have Potential for Reducing Aircraft Greenhouse Gas Emissions, but Challenges Exist

The use of alternative fuels, including those derived from biological sources (biofuels), has the potential to reduce greenhouse gas emissions from aircraft in the future; however, these fuels also present a number of challenges and environmental concerns. While the production and use of biofuels result in greenhouse gas emissions, the extent to which they provide a reduction in greenhouse gas emissions depends on whether their emissions on an energy-content basis are less than those resulting from the production and use of fossil fuels.[41] To date, some assessments of biofuels have shown a potential reduction in greenhouse gas emissions when compared with fossil fuels, such as jet fuel. However, researchers have not agreed on the best approach for determining the greenhouse gas effects of biofuels and the magnitude of any greenhouse gas reductions attributable to their production and use.[42] FAA, EPA, and U.S. Air Force officials we met with said that quantifying the life-cycle emission of biofuels is difficult, but work in this area is currently under way.[43] For example, according to EPA, the agency has developed a comprehensive methodology to determine the life-cycle emissions, including both direct and indirect emissions, of a range of biofuels. This methodology, which involved extensive coordination with experts outside of and across the federal government, was included in the recent notice of proposed rulemaking on the renewable fuel standard. Non-oil-energy sources, such as hydrogen, have potential for providing energy for ground transport, but many experts we met with said that such sources are unlikely to have use for commercial aircraft given technological, cost, and potential safety issues.[44]

According to experts we interviewed, a variety of sources could be used to produce biofuels for aircraft, including biomasses such as switchgrass and forest and municipal waste; and oils from jatropha (a drought-resistant plant that can grow in marginal soil), algae, camelina (a member of the mustard family that can grow in semiarid regions), palm, and soy. However, many experts claim that some of these crops are unsuitable for use as biofuels because they may have negative environmental and economic consequences, such as potentially reducing the supply and quality of water, reducing air quality and biodiversity, and limiting global food supplies. For example, cultivating palm for biofuel production might lead to deforestation, thereby increasing both greenhouse gas emissions and habitat loss. In addition, jatropha has been identified as an invasive species in some regions and, because of its aggressive growth, may have the potential to reduce available habitat for native species. According to experts we met with, algae, on the other hand, are seen as a potentially viable source: they can be grown using saltwater and in a variety of other environments. In addition, according to DOT, camelina appears to be a potential biofuel source in the short term as it is not currently used for food and uses limited water for development.

However, many experts we interviewed raised questions about the availability of future supplies of biofuels. According to the experts, large investments in fuel production facilities will likely be needed because little industrial capacity and compatible infrastructure currently exist to create biofuels.[45] The cost of current algae conversion technology has, for example raised obstacles to the commercial-scale production needed to obtain significant supplies in the future. Given that future alternative fuels will have many uses, airlines will compete with other sources, including road transportation, for those limited supplies. Compared with the market for ground transport, the market for fuels for commercial aviation is small, leading some experts to believe that fuel companies are more likely to focus their biofuel efforts on the ground transport market than on the commercial aviation market. Some experts we met with said that given the relatively small size of the market, limited biofuel supplies should be devoted to road transportation since road transportation is the largest contributor of emissions from the transportation sector.[46]

A large number of industry and government participants, including airlines, fuel producers, and manufacturers, are currently conducting research and development on alternative fuels for aircraft. One effort is the Commercial Aviation Alternative Fuels Initiative, whose members include FAA, airlines, airports, and manufacturers. The goal of this initiative is to "promote the development of alternative fuels that offer equivalent levels of safety and compare favorably with petroleum-based jet fuel on cost and environmental bases, with the specific goal of enhancing security of energy supply." Any developed biofuel will be subject to the same certification as petroleum-based jet fuel to help ensure its safety. In addition, other government efforts are under way, most notably the Biomass Research and Development Initiative. This initiative is a multiagency effort to coordinate and accelerate all federal biobased products and bioenergy research and development. The Department of Transportation is one of the initiative's participants.

Finally, the aviation industry has conducted a number of test flights using a mixture of biofuels and jet fuel. These test flights have demonstrated that fuel blends containing biofuels have potential for use in commercial aircraft. In February 2008, Virgin Atlantic Airlines conducted a demonstration flight of a Boeing 747 fueled by a blend of jet fuel (80 percent) and coconut- and babassu-oil-based fuels (20 percent). In December 2008, Air New Zealand conducted a test flight of a Boeing 747 fueled by a blend consisting of an equal mixture of jet fuel and jatropha oil. In January 2009, Continental Airlines conducted a test flight using a fuel blend of 50 percent jet fuel, and a jatropha and algae biofuel blend on a Boeing 737.[47] In January 2009, Japan Airlines conducted a test flight of a Boeing 747 fueled by a blend including camelina oil. According to the airlines, the results of all these tests indicate that there was no change in performance when engines were fueled using the biofuel blends. For example, the pilot of the Air New Zealand test flight noted that both on- ground and in-flight tests indicated that the aircraft engines performed well while using the biofuel.

Improvements to Reduce Emissions from Aircraft Face Challenges and According to Experts Adopting Them May Not Be Enough to Offset Future Market and Emissions Growth

Future fuel prices are likely to be a major factor in influencing the development of low-emissions technologies for commercial aviation. According to the airline industry, fuel costs provide an incentive for airlines to reduce fuel consumption and emissions. However, according to some experts we interviewed, short-term increases in fuel prices may not provide enough of an incentive for the industry to adopt certain low- emission improvements. For example, the commercial airlines would have greater incentive to adopt fuel saving technologies if the projected fuel savings are greater than the improvement's additional life-cycle cost. The higher existing and projected fuel prices are, the more likely airlines would undertake such improvements, all else the same. One expert said that if fuel costs were expected to consistently exceed $140 per barrel in the future, much more effort would be made to develop a finished open rotor engine quickly. The price of fuel as a factor in providing an incentive for the development and adoption of low-emission technologies is seen in some historical examples in NASA research. While winglets were first developed through a NASA research program in the 1970s, they were not used commercially until a few years ago when higher fuel prices justified their price. Additionally, although NASA currently is sponsoring research into open rotor engines, the agency also did so in the 1980s in response to high fuel prices. That research was discontinued before the technology could be matured, however, when fuel prices dropped dramatically in the late 1980s.

In addition, the current economic recession has impacted commercial airlines and may cause some airlines to cut back on purchases of newer and more fuel-efficient aircraft. For example, the U.S. airline industry lost about $3.7 billion in 2008, and while analysts are uncertain about its profitability 2009, some analysts predict industry profits of around $4 billion to $10 billion. In addition, Boeing has reported a number of recent cancellations of orders for the fuel-efficient 787 Dreamliner. According to one expert we met with, when airlines are low on cash, they are unlikely to undertake improvements that will reduce their fuel consumption and emissions, even if the savings from fuel reductions will ultimately be greater than the cost of the improvement because they have so little cash. This expert said, for example, that although it may make financial sense for airlines to engage in additional nonsafety-related engine maintenance to reduce fuel burn and emissions, they may not do so because they lack sufficient cash.

Although some airlines may adopt technologies to reduce their future emissions, these efforts may not be enough to mitigate the expected growth in air traffic and related increase in overall emissions through 2050. Although IPCC's forecast, as mentioned earlier, assumes future technological improvements leading to annual improvements in fuel efficiency, it excludes or doesn't account for the possibility that some airlines might adopt biofuels or other potential breakthrough technologies. Nonetheless, even if airlines adopt such technologies, some experts believe that emissions will still be higher in 2050 under certain conditions than they were in 2000. One expert we met with did a rough estimate of future emissions from aircraft assuming the adoption of many low-carbon technologies such as blended wing-body, operational improvements, and biofuels. He used IPCC's midrange forecast of emissions to 2050 as a baseline for future traffic and found that even assuming the introduction of these technologies, global emissions in 2050 would continue to exceed 2000 emissions levels.[48] Had a lower baseline of emissions been used, forecasted emissions may have been lower. He acknowledged that more work needs to be done in this area. Another study by a German research organization modeled future emissions assuming the adoption of technological improvements, as well as biofuels, to reduce emissions. This study assumed future traffic growth averaging 4.8 percent between 2006 and 2026 and 2.6 percent between 2027 and 2050.[49] While this study forecasted improvements in emissions relative to expected market growth, it estimated that by 2050 total emissions would still remain greater than 2000 emissions levels.

GOVERNMENTS CAN USE A VARIETY OF POLICY OPTIONS TO HELP REDUCE COMMERCIAL AIRCRAFT EMISSIONS, BUT THE COSTS AND BENEFITS OF EACH VARY

Governments have a number of policy options—including policies that set a price on emissions, market-based measures like a cap-and-trade program or a tax, regulatory standards, and funding for research and development—they could use to help reduce greenhouse gas emissions from commercial aviation and other sectors of the economy. The social benefits (for example, resulting from emissions reductions) and costs associated with each option vary, and the policies may affect industries and consumers differently. However, economic research indicates that market-based policies are more likely to better balance the benefits and costs of achieving reductions in greenhouse gases and other emissions (or, in other words, to be more economically efficient). In addition, research and development spending could complement market-based measures or standards to help facilitate the development and deployment of low-emissions technologies. However, given the relatively small current and forecasted percentage of global emissions generated by the aviation sector, actions taken to reduce aviation emissions alone, and not emissions from other sectors, could be costly and have little potential impact on reducing global greenhouse gas emissions.[50]

Market-Based Policies Could Be Used to Provide Airlines and Other Sources with an Economic Incentive to Reduce Greenhouse Gas Emissions

Economists and other experts we interviewed stated that establishing a price on greenhouse gas emissions through market-based policies, such as a cap-and-trade program or a tax on emissions from commercial aircraft and other sources, would provide these sources with an economic incentive to reduce their emissions. Generally, a cap-and-trade program or an emissions tax (for example, on carbon dioxide) can achieve emissions reductions at less cost than other policies because they would give firms and consumers the flexibility to decide when and how to reduce their emissions. Many experts we surveyed said that establishing a price on emissions through a cap-and-trade program or a tax would help promote the development and adoption of a number of low-emissions technologies for airlines, including open rotor engines and blended wing-body aircraft. Another market-based policy, subsidy programs, such as a payment per unit of emissions reduction, can in principle provide incentives for firms and consumers to reduce their greenhouse gas emissions. However, subsidy programs need to be financed—for example through existing taxes or by raising taxes—and can create perverse incentives resulting in higher emissions.

Cap-and-Trade Program

One market-based option for controlling emissions is a cap-and-trade program. Also known as an emissions trading program, a cap-and-trade program would limit the total amount of emissions from regulated sources. These sources would receive, from the government, allowances to emit up to a specific limit—the "cap." The government could sell the allowances through an auction or provide them free of charge (or some combination of the two). In addition, the government would establish a market under which the regulated sources could buy and sell allowances with one another. Sources that can reduce emissions at the lowest cost could sell their allowances to other sources with higher emissions reduction costs. In this way, the market would establish an allowance price, which would represent the price of carbon dioxide (or other greenhouse gas) emissions. Generally, according to economists, by allowing sources to trade allowances, policy makers can achieve emissions reductions at the lowest cost.

A cap-and-trade program can be designed to cap emissions at different points in the economy. For example, a cap-and-trade program could be designed to cap "upstream" sources like fuel processors, extractors, and importers. Under this approach, a cap would be set on the emissions potential that is inherent in the fossil fuel. The upstream cap would restrain the supply and increase the prices of fossil fuels and thus the price of jet fuel relative to less carbon-intensive alternatives. Alternatively, under a "downstream" program, direct emitters, such as commercial airlines, would be required to hold allowances equal to their total carbon emissions each year. (See fig. 9.) However, economic research indicates that both types of programs would provide commercial airlines with an incentive to reduce their fuel consumption in the most cost-effective way for each airline, such as by reducing weight, consolidating flights, or using more fuel-efficient aircraft, if they were included in such a program. To the extent that airlines would pass along any program costs to customers through higher passenger fares and shipping rates, travelers and shippers could respond in various ways, including by traveling less frequently or using a different, cheaper transportation mode.[51]

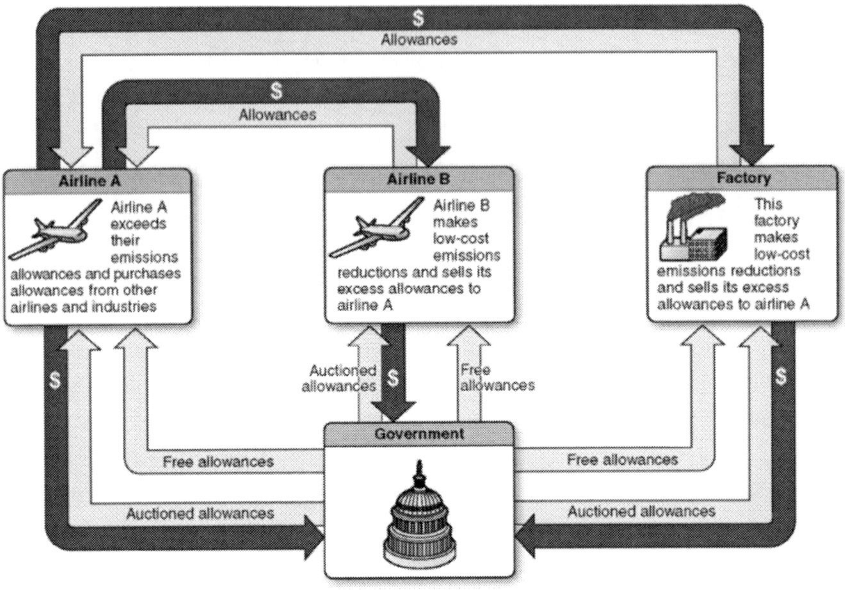

Source: Gao

Figure 9. A Potential Cap-and-Trade Program Regulating Airlines and Other Emissions Sources

The effectiveness of a cap-and-trade program in balancing the benefits and costs of the emission reductions could depend on factors included in its design. Generally, by establishing an upper limit on total emissions from regulated sources, a cap-and-trade program can provide greater certainty than other policies (for example, an emissions tax) that emissions will be reduced to the desired level. Regulated sources would be required to hold allowances equal to their total emissions, regardless of the cost. However, allowance prices could be volatile, depending on factors such as changes in energy prices, available technologies, and weather,[52] making it more expensive for sources to meet the cap. To limit price volatility, a cost- containment mechanism called a “safety valve” could be incorporated into the cap-and-trade program to establish a ceiling on the price of allowances. For example, if allowance prices rose to the safety-valve price, the government could sell regulated sources as many allowances as they would like to buy at the safety-valve price.[53] Although the safety valve could limit price spikes, the emissions cap would be exceeded if the safety valve were triggered.

In addition, the baseline that is used to project future emissions and set the emissions cap can affect the extent to which a cap-and-trade program will contain or reduce emissions.[54] The point in time on which a baseline is set also can influence the environmental benefits of a cap-and-trade program. For example, some environmental interest groups in Europe have claimed that the environmental benefits of including aviation in the EU ETS will be minimal, since the emissions cap will be based on the mean average of aviation emissions from 2004 through 2006, leading to minimal future emissions reductions.[55]

In addition, industry groups and other experts have raised concerns that a cap-and-trade program could be administratively burdensome to the government, which would need to determine how to allocate the allowances to sources, oversee allowance trading, and monitor and enforce compliance with the program. Generally speaking, an upstream program may have lower administrative costs than a downstream program because it would likely involve fewer emissions sources.

Some members of the aviation industry have said they view open and global cap-and-trade programs positively, although they report that not all types of cap-and-trade programs will work for them. For instance, ICAO and other industry organizations have said they would prefer an open capand-trade program (in which airlines are allowed to trade allowances with other sectors and sources) to a closed one (in which airlines are allowed to trade emissions allowances only with one another) because an open program would give airlines more flexibility in meeting their emissions cap. Staff we met with at the Association of European Airlines expressed willingness for aviation to participate in a cap-and-trade program as long as it is global in scope, is an open system, is not in addition to similar taxes, and does not double-count emissions.[56] In addition, some industry groups and government agencies we met with said that a global program would best ensure that all airlines would take part in reducing emissions.

Cap-and-Trade Plans and Legislation

Some countries are planning to address aviation emissions through cap-and-trade programs. The European Union originally implemented the EU ETS in 2005, covering industries representing about 50 percent of its carbon dioxide emissions.[57] The EU is planning on including all covered flights by aircraft operators flying into or out of EU airports, starting in 2012.[58] Please see appendix I for more details on the EU ETS, including a comprehensive discussion of the potential legal implications and stakeholders' positions on this new framework. Other countries are considering cap-and-trade programs that would affect the aviation sector.[59]

In addition, the United States is currently considering and has previously considered cap-and-trade programs:

- H.R. 2454, the American Clean Energy and Security Act of 2009, 111th Cong. (2009), would create a cap-and-trade program for greenhouse gas emissions for entities responsible for 85 percent of emissions in the United States. The current language proposes to regulate producers and importers of any petroleum-based liquid fuel, including aircraft fuel, as well as other entities, and calls for an emissions cap in 2050 that would be 83 percent lower than 2005 emissions. The bill also calls for the emissions cap in 2012 to be 3 percent below 2005 levels, and in 2020 to be 20 percent below 2005 levels. In addition, the Obama Administration's fiscal year 2010 budget calls for the implementation of a cap-and-trade program to regulate emissions in the United States. The budget calls for emissions reductions so that emissions in 2020 are 14 percent below 2005 levels and emissions in 2050 are 83 percent below 2005 levels.
- Additionally in this Congress, the Cap and Dividend Act,[60] also proposes a cap-and-trade program for carbon dioxide emissions beginning in 2012, which would include jet fuel emissions. This program's covered entities would include entities that would make the first sale in U.S. markets of oil or a derivative product used as a combustible fuel, including jet fuel. The bill would require the Secretary of the Treasury, in consultation with the EPA Administrator, to establish the program's emission caps in accordance with the following targets: the 2012 cap would equal 2005 emissions; the 2020 cap would equal 75 percent of 2005 emissions; the 2030 cap would equal 55 percent of 2005 emissions; the 2040 cap would equal 35 percent of 2005 emissions; and the 2050 cap would equal 15 percent of 2005 emissions.
- A number of bills creating a cap-and-trade program also were introduced in the 110th Congress but did not pass. For example, a bill sponsored by Senators Boxer, Warner, and Lieberman would have established a cap-andtrade program that covered petroleum refiners and importers, among other entities.[61] The costs of the regulation would have been borne by these refiners and importers who would likely have passed on those costs to airlines through increases in the price of jet fuel.

Emissions Taxes

An emissions tax is another market-based policy that could be used to reduce emissions from commercial aviation and other emissions sources. Under a tax on carbon dioxide (or other greenhouse gas), the government would levy a fee for every ton of carbon dioxide emitted. Similar to a cap-and-trade program, a tax would provide a price signal to commercial airlines and other emission sources, creating an economic incentive for them to reduce their emissions. A carbon tax could be applied to "upstream" sources such as fuel producers, which may in turn pass along the tax in the form of higher prices to fuel purchasers, including commercial airlines. Similar to a cap-and-trade program, emissions taxes would provide regulated sources including commercial airlines with an incentive to reduce emissions in the most cost-effective way, which might include reducing weight, consolidating flights, or using more fuel-efficient aircraft.

According to economic theory, an emissions tax should be set at a level that represents the social cost of the emissions.[62] Nonetheless, estimates of the social costs associated with greenhouse gas emissions vary. For example, IPCC reported that the social costs of damages associated with greenhouse gas emissions average about $12 per metric ton[63] of carbon dioxide (in 2005 dollars) with a range of $3 to $95 per ton (in 2005 dollars).

Economic research indicates that an emissions tax is generally a more economically efficient policy tool to address greenhouse gas emissions than other policies, including a cap-and-trade program, because it would better balance the social benefits and costs associated with the emissions reductions. In addition, compared to a cap-and-trade program, an emissions tax would provide greater certainty as to the price of emissions. However, it would in concept provide less certainty about emissions reductions because the reductions would depend on the level of the tax and how firms and consumers respond to the tax.[64]

Subsidies

Subsidies are another market-based instrument that could, in principle, provide incentives for sources to reduce their emissions. For example, experts we met with said that the government could use subsidies to encourage industry and others to adopt existing low-emissions technologies and improvements, such as winglets. In addition, some experts told us that NextGen-related technologies are candidates for subsidies because of the high costs of the technologies and the benefits that they will provide to the national airspace system. According to IPCC, subsidies can encourage the diffusion of new low-emissions technologies and can effectively reduce emissions. For example, as newer, more fuel- efficient engines are developed and become commercially available, subsidies or tax credits could lower their relative costs and encourage airlines to purchase them.

Although subsidies are similar to taxes, economic research indicates that some subsidy programs can be economically inefficient, and need to be financed (for example, using current tax revenue or by raising taxes). For example, although some subsidy programs could lead to emissions reductions from individual sources, they may also result in an overall increase by encouraging some firms to remain in business longer than they would have under other policies such as an emissions tax.

Distribution of Costs under Market-based Measures

Both a cap-and-trade program and an emissions tax would impose costs on the aviation sector and other users of carbon-based fuels. The extent to which the costs associated with an emissions control program are incurred by commercial airlines and passed on will depend on a number of economic factors, such as the level of market competition and the responsiveness of passengers to changes in price. Officials of some industry organizations we met with said that because airlines are in a competitive industry with a high elasticity of demand,[65] they are constrained in passing on their costs, and the costs to industry likely will be large. The Association of European Airlines reported that airlines will have very limited ability to pass on the costs of the EU ETS. Furthermore, the International Air Transport Association has estimated that the costs to the industry of complying with the EU ETS will be €3.5 billion in 2012,[66] with annual costs subsequently increasing.[67] Others we interviewed, however, stated that airlines will be able to pass on costs, and the increases in ticket prices will not be large. For example, the EU estimates that airlines will be able to pass on most of the costs of their compliance with the EU ETS, which will result in an average ticket price increase of €9 on a medium-haul flight.[68] However, the revenue generated by the tax or by auctioning allowances could be used to lessen the overall impact on the economy, or the impact on certain groups (for example, low income) or sectors of the economy by, for example, reducing other taxes.[69]

Finally, according to some airline industry representatives, a program to control greenhouse gas emissions would add to the financial burden the aviation industry and its consumers already face with respect to other taxes and fees. For example, passenger tickets in the United States are subject to a federal passenger ticket tax of 7.5 percent, a segment charge of $3.40 per flight segment, and fees for security and airport facilities (up to $4.50 per airport). In addition, international flights are subject to departure taxes and customs-related fees. However, none of these taxes and fees attempt to account for the cost of greenhouse gas emissions, as a tax or cap-and-trade program would do. In addition, the revenue generated from an emissions tax or by auctioning allowances under a cap-and-trade program, could be used to offset other taxes, thereby lessening the economic impact of the program.

EMISSIONS STANDARDS COULD LIMIT EMISSIONS FROM SPECIFIC TECHNOLOGIES, BUT ARE GENERALLY NOT AN ECONOMICALLY EFFICIENT APPROACH FOR REDUCING GREENHOUSE GAS EMISSIONS

Mandating the use of certain technologies or placing emissions limits on aircraft and aircraft engines are also potential options for governments to address aircraft emissions. Standards include both technology standards, which mandate a specific control technology such as a particular fuel- efficient engine, and performance standards, which may require polluters to meet an emissions standard using any available method. The flexibility in the performance standards reduces the cost of compliance compared with technology-based standards and, according to DOT, avoids potential aviation safety implications that may occur from forcing a specific technology across a wide range of operations and conditions.

For example, by placing a strict limit on aircraft emissions, a standard would limit the emissions levels from an engine or aircraft. Regulations on specific emissions have been used to achieve specific environmental goals. ICAO's nitrogen oxide standards place limits on nitrogen oxide emissions from newly certified aircraft engines. These standards were first adopted in 1981 and became effective in 1986. Although no government has yet promulgated standards on aircraft carbon dioxide emissions or fuel economy, emissions standards are being discussed within ICAO's Committee on Aviation Environmental Protection and, in December 2007, a number of environmental interest groups filed petitions with EPA asking the agency to promulgate regulations for greenhouse gas emissions from aircraft and aircraft engines. In addition, the American Clean Energy and Security Act of 2009 would require EPA to issue standards for greenhouse gas emissions from new aircraft and new engines used in aircraft by December 31, 2012.[70]

Although standards can be used to limit greenhouse gas emissions levels from aircraft, economic research indicates that they generally are not as economically efficient as market-based instruments because they do not effectively balance the benefits and costs associated with the emissions reductions.[71] For example, unlike market-based instruments, technology standards would give engine manufacturers little choice about how to reduce emissions and may not encourage them to find cost effective ways of controlling emissions.[72] In addition, according to IPCC, because technology standards may require emissions to be reduced in specified ways, they may not provide the flexibility to encourage industry to search for other options for reducing emissions. However, according to EPA, performance standards to address certain emissions from airlines, such as those adopted by ICAO and EPA, gave manufacturers flexibility in deciding which technologies to use to reduce emissions.[73] Nonetheless, although performance standards can provide greater flexibility and therefore be more cost-effective than technology standards, economic research indicates that standards generally provide sources with fewer incentives to reduce emissions beyond what is required for compliance, compared to market-based approaches. Moreover, standards typically apply to new, rather than existing, engines or aircraft, making new engines or aircraft more expensive, and as a result, the higher costs may delay purchases of more fuel-efficient aircraft and engines.

Current international aviation standards also may require international cooperation. Because ICAO sets standards for all international aviation issues, it may be difficult for the U.S. government, or any national government, to set a standard that is not adopted by ICAO, although member states are allowed to do so. Industry groups we met with said that any standards should be set through ICAO and then adopted by the United States and other nations and, as mentioned earlier, some environmental groups have petitioned EPA to set such standards.

Government-Sponsored Research and Development Can Help Encourage the Development and Adoption of Low-Emissions Technologies, but May Be Costly to Governments

Government-sponsored research into low-fuel consumption and low-emissions technologies can help foster the development of such technologies, particularly in combination with a tax or a cap-and-trade program. Experts we surveyed said that increased government research and development could be used to encourage a number of low-emissions technologies, including open rotor engines and blended wing-body aircraft. According to the *Final Report of the Commission on the Future of the United States Aerospace Industry*, issued in 2002, the lack of longterm investments in aerospace research is inhibiting innovation in the industry and economic growth. This study also asserted that national research and development on aircraft emissions is small when compared with the magnitude of the problem and the potential payoffs that research drives. Experts we met with said that government sponsorship is crucial, especially for long-term fundamental research, because private companies may not have a sufficiently long-term perspective to engage in research that will result in products for multiple decades into the future. According to one expert we interviewed, the return on investment is too far off into the future to make it worthwhile for private companies. NASA officials said that private industry generally focuses only on what NASA deems the "next generation conventional tube and wing technologies," which are usually projected no more than 20 years into the future. Furthermore, raising fuel prices or placing a price on emissions through a tax or capand-trade program is likely to encourage greater research by both the public and private sectors into low-emissions technologies because it increases the pay off associated with developing such technologies.

Various U.S. federal agencies, including NASA and FAA, have long been involved in research involving low-emissions technologies.[74] For example, NASA's subsonic fixed-wing research program is devoted to the development of technologies that increase aircraft performance, as well as reduce both noise levels and fuel burn. Through this program, NASA is researching a number of different technologies to achieve those goals, including propulsion, lightweight materials, and drag reduction. The subsonic fixed-wing program is looking to develop three generations of aircraft with increasing degrees in technology development and fuel burn improvements—the next-generation conventional tube and wing aircraft, the unconventional hybrid wing-body aircraft, and advanced aircraft concepts.[75] NASA follows goals set by the *National Plan for Aeronautics Research and Development and Related Infrastructure* for fuel efficiency improvements[76] for each of these generations (see table 6).[77]

However, budget issues may affect NASA's research schedule. As we have reported, NASA's budget for aeronautics research was cut by about half in the decade leading up to fiscal year 2007, when the budget was $717 million.[78] Furthermore, NASA's proposed fiscal year 2010 budget calls for significant cuts in aeronautics research, with a budget of $569 million. As NASA's aeronautics budget has declined, it has focused more on fundamental research and less on demonstration work. However, as we have reported, NASA and other officials and experts agree that federal research and development efforts are an effective means of achieving emissions reductions in the longer term.[79] According to NASA officials, the research budget for its subsonic fixed-wing research program, much of which is devoted to technologies to reduce emissions and improve fuel efficiency, will be about $69 million in 2009.

FAA has proposed creating a new research consortium to focus on emissions and other issues. Specifically, FAA has proposed the Consortium for Lower Energy, Emissions, and Noise, which would fund, on a 50-50 cost share basis with private partners, research and advanced development into low-emissions and low-noise technologies, including alternative fuels, over 5 years. FAA plans that the consortium will mature technologies to levels that facilitate uptake by the aviation industry. The consortium contributes to the goal set by the National Plan for Aeronautics, Research and Development and Related Infrastructure to reduce fuel burn by 33 percent compared with current technologies. The House FAA Reauthorization Bill (H.R. 915, 111th Cong. (2009)) would provide up to $108 million in funding for the consortium for fiscal years 2010 through 2012.

Lastly, the EU has two major efforts dedicated to reducing aviation emissions. The Advisory Council for Aeronautics Research in Europe (ACARE) is a collaborative group of governments and manufacturers committed to conducting strategic aeronautics research in Europe. According to officials with the European Commission Directorate General of Research, about €150 million to €200 million per year[80] is devoted to basic research through ACARE. Another research effort in Europe is the Clean Sky Joint Technology Initiative, which will provide €1.6 billion[81] over 7 years to fund various demonstration technologies.

Table 6. NASA's Subsonic Fixed-Wing Research Fuel-Reduction Goals

	Research generation		
	Next generation tube and wing	**Unconventional hybrid wing-body**	**Advanced aircraft concepts**
Aircraft fuel burn reduction goal	33 percent fuel reduction (relative to Boeing 737)	40 percent fuel reduction (relative to Boeing 777)	Better than 70 percent fuel reduction
Time frame	2015	2020	2030-2035

Source: NASA.

AGENCY COMMENTS AND OUR EVALUATION

We provided a draft copy of this report to the Department of Defense, the Department of State, the Department of Transportation, the National Aeronautics and Space Administration, and the Environmental Protection Agency for their review.

The Department of Defense had no comments. The Department of State provided comments via email. These comments were technical in nature and we incorporated them as appropriate.

The Department of Transportation provided comments via email. Most of these comments were technical in nature and we incorporated them as appropriate. In addition, DOT stated that our statements indicating that the use of future technological and operational improvements may not be enough to offset expected emissions growth is not accurate given the potential adoption of alternative fuels. We agree that alternative fuels do have potential to reduce aircraft emissions in the future; to the extent that a low-emission (on a life-cycle basis) alternative fuel is available in substantial quantities for the aviation industry, emissions from the aviation industry are likely to be less than they otherwise would be. However, we maintain that given concerns over the potential environmental impacts of alternative fuels, including their life-cycle emissions, as well as the extent to which such fuels are available in adequate supplies at a competitive price, there may be a somewhat limited potential for alternative fuel use to reduce emissions from commercial aircraft in the future, especially the short term. DOT also suggested that we clarify the sources for our discussion about policy options that can be used to address aviation emissions. As much of that discussion is based on economic research and experience with market-based instruments and other policies, we clarified our sources where appropriate.

NASA provided a written response (see app. V) in which they stated that our draft provided an accurate and balanced view of issues relating to aviation and climate change. NASA also provided technical comments that were incorporated as appropriate. EPA provided technical comments via email that were incorporated as appropriate and also provided a written response. (see app. VI).

EPA was concerned that characterizing aircraft emissions standards as being economically inefficient especially compared to market-based measures, might lead readers to believe that emissions standards cannot be designed in a manner that fosters technological innovations and economic efficiency. EPA officials explained that, based on their experience, standards can be designed to optimize technical responses, provide regulated entities with flexibility for compliance and that studies show that EPA regulations have generated benefits in excess of costs. We agree that allowing regulated sources more flexibility in how they meet emissions standards can reduce the costs associated with achieving the emissions reductions. However, economic research indicates that for addressing greenhouse gas emissions, market-based measures such as emissions taxes or cap-and-trade programs would be economically efficient (that is, would maximize net benefits) compared to other approaches, in part because market-based measures can give firms and consumers more flexibility to decide when and how to reduce their emissions. Emissions standards, for example, generally give regulated sources fewer incentives to reduce emissions beyond what is required for compliance. The ultimate choice of what specific policy option or combination of options governments might use and how it should be designed is a complex decision and beyond the scope of our discussion.

Finally, EPA was concerned that our draft report did not adequately discuss the increases in fuel consumption and emissions that have resulted from high rates of market growth and expected continued growth. We believe that our report adequately discusses fuel efficiency as well as fuel consumption and emissions output. In addition, our report discusses that aviation emissions are expected to grow in the long term, despite the potential availability of a number of technological and operational options that can help increase fuel efficiency. In response to this comment, we added additional information on forecasted fuel use by U.S.-based commercial airlines.

We are sending copies of this report to the Secretaries of Defense, State, and Transportation and the Administrators of the Environmental Protection Agency and the National Aeronautics and Space Administration. This report is also available at no charge on the GAO Web site at http://www.gao.gov.

If you or your staffs have any questions concerning this report, please contact me at (202) 512-2834 or flemings@gao.gov. Contact points for our Offices of Congressional Relations and Public Affairs may be found on the last page of this report. Staff members making key contributions to this report are listed in Appendix VII.

Susan A. Fleming

Susan Fleming
Director, Physical Infrastructure Issues

Appendix I. Legal Implications of European Union Emissions Trading Scheme

The European Union's recent decision to include aviation in the European Union's Emissions Trading Scheme (EU ETS), which includes U.S. carriers flying in and out of Europe, is a complex and controversial matter. Preparations by U.S. carriers are already underway for 2012, the first year aircraft operators will be included in the ETS. The inclusion of aviation in the current EU ETS implicates a number of international treaties and agreements and has raised concerns among stakeholders both within and outside the United States. Many stakeholders within the United States have posed that the inclusion of aviation in the ETS violates provisions of these international agreements and is contrary to international resolutions. Others, primarily in Europe, disagree and find aviation's inclusion in the current ETS to be well within the authority set forth in these agreements. In light of these disagreements, the EU may confront a number of hurdles in attempting to include U.S. carriers in the current EU ETS framework.

EU ETS Law

In 2005, the EU implemented its ETS, a cap-and-trade program to control carbon dioxide emissions from various energy and industrial sectors. On December 20, 2006, the European Commission set forth a legislative proposal to amend the law, or directive, which established the ETS so as to include aviation in the ETS.[82] On July 8, 2008, the European Parliament adopted the legislative resolution of the European Council and on October, 24, 2008, the Council adopted the directive, signaling its final approval.[83] The directive was published in the Official Journal on January 13, 2009, and became effective on February 2, 2009.[84]

Under the amended ETS Directive, beginning on January 1, 2012, a cap will be placed on total carbon dioxide emissions from all covered flights by aircraft operators flying into or out of an EU airport.[85] Emissions will be calculated for the entire flight.[86] For 2012, the cap for all carbon dioxide emissions from covered flights will be set at 97 percent of historical aviation emissions.[87] For the 2013-2020 trading period and subsequent trading periods, the cap will be set to reflect annual emissions equal to 95 percent of historical aviation emissions.[88]

The cap represents the total quantity of emissions allowances available for distribution to aircraft operators. In 2012 and each subsequent trading period, 15 percent of allowances must be auctioned to aircraft operators; the remaining allowances will be distributed to these aircraft operators for free based on a benchmarking process.[89] Individual member states, in accordance with the EU regulation, will conduct the auctions for aircraft operators assigned to that member state.[90] The auction of allowances will be open for anyone to participate. The number of allowances each member state has to auction depends on its proportionate share of the total verified aviation emissions for all member states for a certain year.[91] The member states will be able to use the revenues raised from auctions in accordance with the amended directive.[92] For each trading period, aircraft operators can apply to their assigned member state to receive free allowances. Member states will allocate the free allowances in accordance with a process the European Commission establishes for each trading period.

After the conclusion of each calendar year, aircraft operators must surrender to their assigned member state a number of allowances equal to their total emissions in that year. If an aircraft operator's emissions exceed the number of free allowances it receives, it will be required to purchase additional allowances at auction or on the trading market for EU ETS allowances.[93] In addition, in 2012, aircraft operators will be able to submit certified emissions reductions (CER) and emission reduction units (ERU)—from projects in other countries undertaken pursuant to the Kyoto Protocol's Clean Development Mechanism and Joint Implementation—to cover up to 15 percent of their emissions in lieu of ETS allowances. For subsequent trading periods, aircraft operators' use of CERs and ERUs depends in part on whether a new international agreement on climate change is adopted. However, regardless of whether such an agreement is reached, in the 2013 through 2020 trading period, each aircraft operator will be allowed to use CERs and ERUs to cover at least 1.5 percent of their emissions.[94]

If a country not participating in the EU ETS adopts measures for reducing the climate change impact of flights to participating countries, then the European Commission, in consultation with that country, will consider options to provide for "optimal interaction" between the ETS and that country's regulatory scheme—for example, the Commission may consider excluding from the ETS flights to participating EU ETS countries from that country.[95] Although 2012 is the first year aircraft operators must comply with the ETS law, preparations in the EU[96] and from U.S. carriers[97] began soon after the law went into force.

Legal Implications of the ETS

The inclusion of aviation in the newly amended EU ETS implicates a number of international agreements, policy statements, and a bilateral agreement specific to the United States, including the United Nations Framework Convention on Climate Change (UNFCCC), the Kyoto Protocol to the UNFCCC, the Convention on International Civil Aviation (the 'Chicago Convention'), Resolutions of the International Civil Aviation Organization, and the U.S.-EU Air Transport Agreement (the 'U.S.-EU Open Skies Agreement').

The UNFCCC, a multilateral treaty on global warming that was signed in 1992 and has been ratified by 192 countries, including the United States, seeks to "achieve stabilization of greenhouse gas concentrations in the atmosphere at a level that would prevent dangerous anthropogenic interference with the climate system."[98] Although the UNFCCC required signatory states to formulate a national response to climate change,[99] its mitigation provisions did not require mandatory national emissions targets.[100]

In order to strengthen the commitments articulated in the UNFCCC, the Kyoto Protocol was developed within the UNFCCC's framework and adopted in 1997. The Protocol entered into force in February 2005.[101] The Kyoto Protocol established binding greenhouse gas emissions targets for a number of industrialized nations[102] and the European Economic Community (EEC).[103] Notably, the agreement required these industrialized nations and the EEC to pursue "limitations or reduction of emissions of greenhouse gases ... from aviation ... working through the International Civil Aviation Organization."[104] As of January 2009, 183 countries had ratified the Kyoto Protocol, but not the United States.

Further, the Convention on International Civil Aviation, commonly known as the Chicago Convention, signed on December 7, 1944, sets forth rules on airspace, issues of sovereignty, aircraft licensing and registration, and general international standards and procedures, among others.[105] Notably, the treaty sets forth sovereignty provisions, recognizing that a contracting state has exclusive sovereignty over airspace above its own territory.[106] Provisions potentially applicable to the recent amendment incorporating aviation into the ETS include Article 11,[107] Article 12,[108] Article 15,[109] and Article 24.[110]

Established by the Chicago Convention in 1944, the International Civil Aviation Organization (ICAO) is an agency of the United Nations and is tasked with fostering the planning and development of international aviation. ICAO has issued a number of Assembly Resolutions, which are statements of policy rather than law, including a nonbinding ICAO Resolution A36-22 relating to environmental protection and aviation emissions.[111] This resolution, which supersedes ICAO Resolution A35-5 which had endorsed the further development of an open emissions trading scheme for international aviation, calls for mutual agreement between contracting states before implementation of an emissions trading scheme.[112] Additionally, the Resolution formed a new Group on International Aviation and Climate Change (GIACC) that was tasked with developing and recommending to the ICAO Council a program of action to address international aviation and climate change.[113] GIACC is due to report to the Council later this year.

Finally, the U.S.-EU Air Transport Agreement,[114] signed on April 25 and 30, 2007, and provisionally applied as of March 30, 2008, provided greater flexibility for flights between the United States and the EU, authorizing every U.S. and every EU airline to: operate without restriction on the number of flights, aircraft, and routes, set fares according to market demand; and enter into cooperative arrangements, including codesharing, franchising, and leasing. It includes enhanced opportunities for EU investment in carriers from almost 30 non-EU countries, and enhanced regulatory cooperation in regard to competition law, government subsidies, the environment, consumer protection, and security.[115] Among the provisions potentially applicable to the newly amended EU ETS is Articles 12 relating to charges for use of airports and related facilities and services and Article 3 which prohibits a party from unilaterally limiting service or aircraft type.[116]

Although a number of international agreements, policy statements, and bilateral agreements are in place currently, climate change policies are constantly changing. In December 2009, the Conference of the Parties to the UNFCCC will meet in Copenhagen to discuss and negotiate an "agreed outcome" in order to implement the UNFCCC "up to and beyond 2012."[117]

Stakeholder Positions on Legal Issues

A number of stakeholders have expressed concern as to the legal basis for aviation's inclusion in the EU ETS. In the United States, within the EU community, and in countries throughout the world, public and private entities, as well as legal scholars, have expressed opinions as to whether the inclusion of aviation into the ETS is in compliance with international law.

Stakeholder Positions within the United States

Stakeholders within the United States, such as the executive branch, members of Congress, and the Air Transport Association (ATA), have weighed in on the legality of the newly amended EU ETS which requires compliance by U.S. carriers.

In 2007 and 2008, the executive branch expressed the view that the imposition of the ETS was inconsistent with international law, specifically, the Chicago Convention and the U.S.-EU Air Transport Agreement.[118] While the executive branch has not articulated a position on this issue since mid- 2008, it has expressed the importance of climate change and developing a solution on a global level.[119]

The Air Transport Association (ATA), a trade association representing principle U.S. airlines, also has concluded that the EU ETS's inclusion of aviation violates international law, specifically the Chicago Convention.[120] ATA argues that imposition of the ETS on U.S.-based carriers is contrary to Articles 1, 12, 11,[121] 15[122] and potentially, in the alternative, Article 24.[123] In summary, ATA argues that the ETS, as amended, violates Article 1 and Article 12 provisions of sovereignty and authority. Article 1, which provides contracting states exclusive sovereignty over their airspace,[124] is violated by the EU's extraterritorial reach which covers emissions of nonEU airlines in another states' airspace. Further, Article 12, which requirescontracting states to ensure that aircraft under its jurisdiction are in compliance with rules and regulations relating to the flight and maneuver of aircraft,[125] also is violated. ATA argues that Article 12 gives ICAO primary authority, under the Convention, to set rules for the "flight and maneuver of aircraft" over the "high seas," which precludes the application of rules by one state over the airlines of another state to the extent inconsistent with ICAO rules. Thus, because ICAO has stated that one state can apply emissions trading to the airlines of another state only through mutual consent, ATA contends that the EU's emissions trading coverage of non-agreeing-EU airlines over the high seas is inconsistent with ICAO' s authority.

Additionally, with respect to Article 11, ATA argues that although Article 11 provides authority to states to establish certain rules for admission and departure of aircraft, the authority is limited. States may only establish admission and departure rules consistent with the remainder of the Chicago Convention, which prevents the EU from arguing that Article 11 authorizes EU action. In any event, ATA contends that any rules may only apply "upon entering or departing from or while within the territory of that State," whereas the European scheme reaches outside European territory. Further, ATA finds that the ETS is contrary to Article 15[126] of the Chicago Convention because it imposes a de facto charge for the right to enter or exit an EU member state. In the alternative, ATA argues that there could be a violation of Article 24 of the Convention, which exempts fuel on board an aircraft from duties, fees, and charges. Because the law calculates emissions based on fuel consumption, the purchase of greenhouse gas permits may constitute a "similar ... charge" of fuel on board, according to ATA. Additionally, Article 24 mirrors Article 11 of the U.S.-EU Air Transport Agreement but extends the freedom from taxation/charges on fuel to that purchased in the EU. Thus, ATA argues, the prohibition against the EU levying a fuel tax applies to fuel already on board as well as fuel purchased in the EU.

ATA has publicly expressed harsh opposition to the ETS's inclusion of aviation and has stated that there will be a number of legal challenges from around the globe, including from the United States.[127] ATA has additionally expressed discontent with the newly amended ETS law as a matter of policy, as it siphons money out of aviation which is counterproductive from reinvesting in improving technologies that reduce emissions.[128]

Finally, the Congress is considering the House FAA Reauthorization Bill, H.R. 915, 111th Cong. (2009), which includes an expression of the Sense of the Congress with respect to the newly amended EU ETS.[129] The bill states that the EU's imposition of the ETS, without working through ICAO, is inconsistent with the Chicago Convention, other relevant air service agreements, and "antithetical to building cooperation to address effectively the problem of greenhouse gas emissions by aircraft engaged in international civil aviation."[130] The bill recommends working through ICAO to address these issues.[131]

Stakeholder Positions outside the United States

Stakeholders in the EU community and a not-for-profit business organization have expressed both legal and policy views on the newly amended ETS, as well. An independent contractor for the European Commission's Directorate-General of the Environment (DG Environment) as well as the International Emissions Trading Association (IETA) have both issued opinions in support of aviation's inclusion in the ETS.[132] IETA supports the inclusion of aviation in the EU ETS from a policy perspective, but has not opined on the legality of its inclusion.[133] From a policy standpoint, IETA supports aviation's inclusion on both EU and non-EU carriers so as to share the burden to combat climate change.[134] However, the organization has expressed concerns over a number of issues, some of which include access to project credits, amount of allowances available for auctioning, and allocation calculation.[135]

From a legal perspective, an independent contractor for the DG Environment[136] has issued an opinion supporting aviation's inclusion in the EU ETS.[137] The opinion states that alongside a state's inherent right to enact legislation and the provisions of the EC Treaty for collective action by member states, the legal basis for action is contained in the UNFCCC 's requirement for developed countries to take measures to mitigate climate change, in the Kyoto Protocol which reinforces this position, and under EU law (the EC Treaty), which provides the basis for the EU to act by including aviation in the ETS.[138] Further, the opinion articulates that the ETS does not violate Articles 11,[139] 12,[140] 15,[141] or 24[142] of the Chicago Convention. First, Article 11 is consistent with the newly amended ETS law so long as the "laws and regulations" established do not "discriminate as to nationality of aircraft."[143] Further, Article 12 is inapplicable because emissions trading does not affect the flight or maneuver of aircraft but merely the terms for admission to and departure from EU territory. Additionally, the opinion stated that Article 15 is inapplicable because the coverage of aviation cannot be seen as an "airport charge or similar charge." Even if payment were to occur under an auction system, the allowances are not designed as compensation for the costs of operation and management of airports and air navigation facilities, and consequently, Article 15 is inapplicable.[144] Finally, Article 24 of the Convention does not apply to the Emissions Trading System because trading allowances are "fundamentally different from customs duties."[145] Additionally, the opinion finds policy support for these legal findings in ICAO Resolution A35-5[146] and bilateral air transport agreements.[147]

Additionally, countries outside the European Community have joined the United States in an expression of concerns regarding the imposition of the ETS on non-EU carriers. In an April 2007 letter to the German Ambassado r to the European Union, the United States, Australia, China, Japan, South Korea, and Canada conveyed a "deep concern and strong dissatisfactio n" for the then-proposal to include international civil aviation within the scope of the EU ETS.[148] The letter asks that the EU ETS not include non-EU aircraft unless done by mutual consent.[149] Although supportive of the reduction of greenhouse gas emissions, the ascribing parties argue that the "unilateral" imposition of the ETS on non-EU carriers would potentially violate the Chicago Convention and bilateral aviation agreements with the parties to the letter.[150] Moreover, they write, the proposal runs counter to the international consensus that ICAO should handle matters of international aviation, which was articulated with the ICAO Assembly and the ICAO Council in 2004 and 2006, respectively.[151] The letter closes with a reservation of right to take appropriate measures under international law if the ETS is imposed.[152]

Legal Scholar/Researcher Views

Given the controversial nature and complexity of aviation's inclusion in the EU ETS, a number of scholars in the legal community, both within the United States and the EU, have provided explanatory articles or position papers on the issue of the consistency of the EU's plans with its international legal obligations.[153]

One U.S. law review article by Daniel B. Reagan argues that international aviation emission reductions should be pursued through ICAO given the "political, technical, and legal implications raised by the regulation."[154] This article sets forth that politically, ICAO is the appropriate body because it can work towards uniformity in a complex regulatory arena, incidentally resulting in increased participation from a variety of stakeholders, reduction of resentment, and a reduced likelihood of non-compliance and legal challenges.[155] Further, ICAO has the expertise necessary to technically design aviation's emission reduction regime and is in a position to consider the "economic, political, and technical circumstances of its member states ... ."[156] Finally, Reagan argues that pursuing an emissions reduction regime through ICAO could avoid likely legal challenges which present themselves under the current ETS, as ICAO could facilitate a common understanding of contentious provisions.[157] In conclusion, he proposes that the EU should channel the energy for implementation of the current regime into holding ICAO accountable for fulfilling environmental duties.[158]

In contrast, a law review article published in the European Environmental Law Review in 2007 by Gisbert Schwarze argues that bringing aviation into the EU ETS falls clearly within existing law and is, in fact, mandated.[159] The article presents the case that neither existing traffic rights in member states, bilateral air transport agreements, nor the Chicago Convention pose any legal obstacles.[160] He argues, in fact, that the EU has a mandate under the UNFCCC and the Kyoto Protocol to implement climate change policies which include aviation. First, the article sets forth that the inclusion of aviation does not restrict existing traffic rights or allow or disallow certain aircraft operations in different member states, but rather merely brings the amount of emissions into the decision-making process.[161]

Further, Schwarze explains that imposing the ETS on carriers flying in and out of the EU is well within the Chicago Convention. Article 1 of the

Convention provides contracting states exclusive sovereignty over their airspace which provides the EU with the authority to impose obligations relating to arrival and departures, so long as there is no discrimination on the basis of nationality, as required by Article 11.[162] Additionally, the article sets forth that Article 12, regarding the flight and maneuver of aircraft, is not applicable because, as argued above, the ETS does not regulate certain aircraft operations. Article 15 , which covers charges, is similarly inapplicable because emissions allowances on the free market or through the auctioning process do not constitute a charge.[163] Finally, Article 24 is inapposite as well because the emissions trading system does not constitute a customs duty.[164]

Additionally, Schwarze argues that the bilateral air transport agreements with various nations, such as the Open Skies Agreement with the Unite d States, do not pose any legal barriers to inclusion of aviation in ETS.[165]

These agreements contain a prohibition of discrimination similar to Article 11 of the Chicago Convention and a fair competition clause which require s fair competition among signatories in international aviation as well as prohibits a party from unilaterally limiting traffic. [85] The article argues that so long as the ETS operates without discrimination, it is in conformity with the principle of a sound and economic operation of air services and therefore satisfies the fairness clause.[167] Finally, since the ETS gives only incentive to reduce emissions, it does not regulate the amount of air traffic.[168]

Finally, Schwarze argues that not only is the inclusion of aviation into the EU ETS legally sound, the UNFCCC and Kyoto Protocol mandate its inclusion. The UNFCCC requires all parties to the treaty to adopt national policies and take corresponding measures on the mitigation of climate change consistent with the objective of the convention, recognizin g that this can be done "jointly with other parties."[169] Additionally, the Kyoto Protocol, which sought to strengthen UNFCCC, required Annex 1 parties to pursue "limitations or reduction of emissions of greenhouse gases ... from aviation ... working through the International Civil Aviation Organization."[170] And finally, although not legally binding, ICAO Resolution A35-5 endorses the development of an open emissions trading system for international aviation.[171]

Potential Legal Challenges and Dispute Resolution

Implementation of the new ETS directive will likely face legal challenges before 2012, the first year aircraft operators will be included in the ETS. A number of stakeholders, including ATA, have publicly expressed that there will be "government-to-government legal challenges" and potentially a "multilateral challenge from around the world."[172] Further, countries outside the EU community have joined in support of taking appropriate actions under international law if the ETS is imposed.[173] If challenges are brought forth, they could potentially be brought forth under the Chicago Convention, air service agreements (e.g., U.S.- EU Air Transport Agreement) or potentially in individual member state courts. Each option has its own dispute resolution procedure.

If a challenge is brought forth under the Chicago Convention after failed negotiations, Article 84 of the Convention (Settlement of Disputes) is invoked. Article 84 provides that if there is a disagreement by two or more contracting states which cannot be settled by negotiation, it will be decided upon by the Council.[174] A decision by the Council can be appealed to an agreed-upon ad hoc tribunal or to the Permanent Court of International Justice (now the International Court of Justice) whose decision will be binding.[175]

Air service agreements additionally have dispute resolution procedures and the U.S.-EU Air Transport Agreement is no exception. Article 19 of the U.S.-EU Air Transport Agreement provides that parties to a dispute may submit to binding arbitration through an ad hoc tribunal if negotiatio ns fail.[176] If there is noncompliance with the tribunal's decision and a subsequent agreement between the parties is not reached within 40 days, the other party may suspend the application of comparable ben efits which arise under the agreement.[177]

In addition to bringing legal challenges under international treaties, carriers could potentially mount legal challenges in member states' national courts, according to some legal scholars.[178] Additionally, a European state will potentially be able to take action against a noncompliant carrier under their civil aviation authority in the state's courts.[179]

APPENDIX II. LIST OF EXPERTS

Gerald Bernstein, Stanford Transportation Group
Dennis Bushnell, National Aeronautics and Space Administration
Kenneth Button, George Mason University
Anthony Dean, General Electric
Robert Deering, American Airlines
Christian Dumas, Airbus (France)
Jasper Faber, CE Delft (Netherlands)

Richard Golaszewski, GRA, Incorporated
Preston Henne, Gulfstream Aerospace Corporation
James Hileman, Massachusetts Institute of Technology
Jennifer Holmgren, Universal Oil Products
David Lee, Manchester Metropolitan University (United Kingdom)
Peter Morrell, Cranfield University (United Kingdom)
Andreas Schafer, University of Cambridge (United Kingdom)
Agam Sinha, MITRE Corporation
Julian Tishkoff, United States Air Force
Ian Waitz, Massachusetts Institute of Technology
Donald Wuebbles, University of Illinois

APPENDIX III. DETAILED SURVEY RESULTS

The survey tool used to assess options for reducing commercial aircraft emissions is below, complete with detailed results. We do not include the responses for open-ended questions.

Introduction

This rating tool follows up on our recent interview with you on commercial aviation and greenhouse gas emissions. The tool contains options for reducing commercial aviation emissions that were identified through our interviews with you and other experts, or by GAO. We have placed options in three categories—technologies, operations, and alternative fuels—and are interested in your expert opinion on them. We ask that you rate the options across several factors, providing comments where appropriate.

Instructions for Completing This Tool

You can answer most of the questions by checking boxes or filling in blanks. A few questions request short narrative answers. Please note that these blanks will expand to fit your answer.

- Please use your mouse to navigate throughout the document by clicking on the field or checking the box you wish to fill in. Do not use the “Tab” or “Enter” keys, because doing so may cause formatting problems.

- To select a box, click on it once; to deselect a box, double click on it.

If you prefer, you may print this tool, complete it by hand, and return it by fax. Please use extra paper as necessary to complete the open-ended questions.

Part 1. Technology Options

1. **How would you rate your overall knowledge of technological options to reduce aircraft carbon dioxide (CO_2) emissions, such as aircraft engines and aircraft design technologies, and the costs of those technologies?**

0 None → SKIP TO QUESTION #9
6 Minimal → SKIP TO QUESTION #9
4 Basic → CONTINUE TO QUESTION #2
1 Proficient → CONTINUE TO QUESTION #2
7 Advanced → CONTINUE TO QUESTION #2

2. **In your expert opinion, what is the potential for future fuel savings and CO_2 emissions reductions for the following options?**

	Low potential	Medium potential	High potential	Don’t know
a. Open rotor engines	0	5	8	1
b. Geared Turbo Fan Engines	2	7	3	
c. Distributive Propulsion systems	2	2	5	
d. Lighter airframes (Composites)	2	5	6	
e. Increased Laminar Flow Control	2	6	3	
f. Blended wing-body aircraft	0	5	8	
g. Winglets h.	8	3	3	0
h. Riblets	9	1	0	4

3. **In your expert opinion, what would be the potential R&D costs to develop the following options for commercial use?**

	Lowcosts	Medium Cost	High Cost	Don't know
a. Open rotor engines	1	6	6	1
b. Geared Turbo Fan Engines	3	7	3	1
c. Distributive Propulsion systems	0	1	7	6
d. Lighter airframes (Composites)	1	6	6	1
e. Increased Laminar Flow Control	0	4	7	3
f. Blended wing-body aircraft	1	1	12	0
g. Winglets	14	0	0	0
h. Riblets	7	3	0	4

4. **Given your answer to question two, what would be the potential costs to the air transport industry to procure, operate and maintain the following options to achieve those fuel savings and CO_2 emissions reductions?**

	Lowcosts	Medium Cost	High Cost	Don't know
a. Open rotor engines	2	7	3	2
b. Geared Turbo Fan Engines	2	9	1	2
c. Distributive Propulsion systems	0	2	6	6
d. Lighter airframes (Composites)	3	6	3	2
e. Increased Laminar Flow Control	1	4	5	4
f. Blended wing-body aircraft	2	3	9	0
g. Winglets	14	0	0	0
h. Riblets	9	1	0	4

5. **In your expert opinion, what is the level of public acceptance for the following conceptual options,**

	Low acceptance	Medium acceptance	High acceptance	Don't know
a. Open rotor engines	6	7	0	1
b. Geared Turbo Fan Engines	0	2	11	1
c. Distributive Propulsion systems	2	2	4	6
d. Lighter airframes (Composites)	0	3	9	2
e. Increased Laminar Flow Control	0	4	8	2
f. Blended wing-body aircraft	4	5	2	3
g. Winglets	0	0	13	1
h. Riblets	0	1	9	4

6. In your expert opinion, given our best knowledge about future market conditions, and absent government intervention, how long would it take for the private sector to adopt these technologies?

	Short time frame (< 5 years)	**Medium time frame (5 – 15 years)**	**Long time frame (> 15 years)**	**Never**	**Don't know**
a. Open rotor engines	1	9	3	0	1
b. Geared Turbo Fan Engines	6	6	0	0	2
c. Distributive Propulsion systems	0	0	6	2	6
d. Lighter airframes (Composites)	3	9	1	0	1
e. Increased Laminar Flow Control	2	4	5	0	3
f. Blended wing-body aircraft	0	2	10	2	0
g. Winglets	14	0	0	0	0
h. Riblets	5	3	1	0	5a

7. What major challenges exist to widespread use of ...

a. Open rotor engines	
b. Geared Turbo Fan Engines	
c. Distributive Propulsion systems	
d. Lighter airframes (Composites)	
e. Increased Laminar Flow Control	
f. Blended wing-body aircraft	
g. Winglets	
h. Riblets	

8. What actions could the U.S. federal government undertake to promote the development and/or adoption of ...

a. Open rotor engines	
b. Geared Turbo Fan Engines	
c. Distributive Propulsion systems	
d. Lighter airframes (Composites)	
e. Increased Laminar Flow Control	
f. Blended wing-body aircraft	
g. Winglets	
h. Riblets	

Part 2. Operational Options

9. How would you rate your overall knowledge of operational options to reduce aircraft fuel usage and CO_2 emissions, such as air traffic management and airline operations?

0 None → SKIP TO QUESTION #17
5 Minimal → SKIP TO QUESTION #17
7 Basic → CONTINUE TO QUESTION #10
4 Proficient → CONTINUE TO QUESTION #10
2 Advanced → CONTINUE TO QUESTION #10

10. In your expert opinion, what is the future potential for fuel savings and CO_2 emissions reductions for the following options?

	Low potential	**Medium potential**	**High potential**	**Don't know**
a. Reduction of on-board weight	7	4	2	0
b. Engine washing	8	4	0	1
c. Limited use of paint on airframes	9	3	0	1
d. Increased engine maintenance	8	3	1	1
e. Single engine taxing	7	6	0	0
f. Use of APU on ground at gate	10	2	0	1
g. Automatic Dependent Surveillance – Broadcast (ADS-B)	2	6	1	4
h. Required Navigation Performance (RNP)	2	4	2	5
i. Continuous Descent Arrivals (CDA)	5	5	2	1
j. NEXTGEN Weather	3	5	1	4
k. Enhanced vision systems	6	3	0	4
l. Synthetic vision systems	6	2	0	5
m. Reduced Vertical Separation	3	8	1	1
n. Flying at slower speeds	5	5	3	0
o. Formation flying	5	2	2	4
p. Air to air refueling	7	1	0	5
q. Reduced flight frequency with larger aircraft	6	3	3	1
r. Multi-stage long distance travel	6	4	1	2

11. In your expert opinion, what would be the potential R&D costs to develop the following options for commercial use?

	Low Costs	Medium Costs	High Costs	Don't know
a. Reduction of on-board weight	8	4	1	0
b. Engine washing	11	0	0	2
c. Limited use of paint on airframes	13	0	0	0
d. Increased engine maintenance	11	2	0	0
e. Single engine taxing	13	0	0	0
f. Use of APU on ground at gate	11	0	0	2
g. Automatic Dependent Surveillance – Broadcast (ADS-B)	1	6	1	5
h. Required Navigation Performance (RNP)	3	4	0	6
i. Continuous Descent Arrivals (CDA)	4	7	0	2
j. NEXTGEN Weather	1	4	3	5
k. Enhanced vision systems	2	5	1	5
l. Synthetic vision systems	2	5	1	5
m. Reduced Vertical Separation	9	1	1	2
n. Flying at slower speeds	13	0	0	0
o. Formation flying	3	3	4	3
p. Air to air refueling	0	3	5	5
q. Reduced flight frequency with larger aircraft	1 2	0	1	0
r. Multi-stage long distance travel	11	0	1	1

12. Given your answer to question ten, what would be the potential costs to the air transport industry to adopt the following options to achieve those fuel savings and CO_2 emissions reductions?

	Low costs	Medium costs	High costs	Don't know
a. Reduction of on-board weight	8	5	0	0
b. Engine washing	8	4	0	1
c. Limited use of paint on airframes	11	2	0	0
d. Increased engine maintenance	5	6	1	1
e. Single engine taxing	13	0	0	0
f. Use of APU on ground at gate	10	1	0	2
g. Automatic Dependent Surveillance –	2	4	2	5

Broadcast (ADS-B)				
h. Required Navigation Performance (RNP)	4	3	0	6
i. Continuous Descent Arrivals (CDA)	8	3	0	1
j. NEXTGEN Weather	2	3	1	6
k. Enhanced vision systems	1	5	2	5
l. Synthetic vision systems	1	6	1	5
m. Reduced Vertical Separation	9	2	0	2
n. Flying at slower speeds	11	2	0	0
o. Formation flying	2	4	2	5
p. Air to air refueling	0	4	5	4
q. Reduced flight frequency with larger aircraft	9	1	3	0
r. Multi-stage long distance travel	5	4	2	2

13. In your expert opinion, what is the level of public acceptance for the following options?

	Low acceptance	Medium acceptance	High acceptance	Don't know
Reduction of on-board weight	0	5	8	0
Engine washing	0	0	11	2
Limited use of paint on airframes	1	2	10	0
Increased engine maintenance	0	0	12	1
Single engine taxing	0	2	11	0
Use of APU on ground at gate	0	2	9	2
Automatic Dependent Surveillance – Broadcast (ADS-B)	0	0	9	4
Required Navigation Performance (RNP)	0	0	8	5
Continuous Descent Arrivals (CDA)	0	1	12	0
NEXTGEN Weather	0	1	8	4
Enhanced vision systems	0	1	9	3
Synthetic vision systems	1	1	8	3
Reduced Vertical Separation	1	6	6	0
Flying at slower speeds	7	5	1	0
Formation flying	8	3	1	1
Air to air refueling	10	1	0	2
Reduced flight frequency with larger aircraft	7	5	1	0
Multi-stage long distance travel	8	3	0	1

14. In your expert opinion, given our best knowledge about future market conditions, and absent government intervention, how long would it take for tht e private sector to adopt these technologies?

	Short timeframe (> 15 years)	Medium timeframe (5-15 years)	Long timeframe (> 15 years)	Never	Don't know
a. Reduction of on-board weight	8	5	0	0	0
b. Engine washing	11	1	0	0	1
c. Limited use of paint on airframes	9	4	0	0	0
d. Increased engine maintenance	12	1	0	0	0
e. Single engine taxing	13	0	0	0	0
f. Use of APU on ground at gate	8	3	0	0	2
g. Automatic Dependent Surveillance – Broadcast (ADS-B)	3	5	0	0	5
h. Required Navigation Performance (RNP)	6	1	0	0	6
i. Continuous Descent Arrivals (CDA)	7	5	0	0	1
j. NEXTGEN Weather	2	5	0	0	6
k. Enhanced vision systems	2	5	1	0	5
l. Synthetic vision systems	1	4	3	0	5
m. Reduced Vertical Separation	5	6	1	0	1
n. Flying at slower speeds	11	0	2	0	0
o. Formation flying	2	4	4	2	1
p. Air to air refueling	0	1	8	1	3
q. Reduced flight frequency with larger aircraft	8	3	1	1	0
r. Multi-stage long distance travel	6	1	2	2	2

15. What major challenges exist to widespread use of ...

a. Reduction of on-board weight	
b. Engine washing	
c. Limited use of paint on airframes	
d. Increased engine maintenance	
e. Single engine taxing	
f. Use of APU on ground at gate	
g. Automatic Dependent Surveillance – Broadcast (ADS-B)	
h. Required Navigation Performance (RNP)	
i. Continuous Descent Arrivals (CDA)	
j. NEXTGEN Weather	
k. Enhanced vision systems	
l. Synthetic vision systems	

m. Reduced Vertical Separation	
n. Flying at slower speeds	
o. Formation flying	
p. Air to air refueling	
q. Reduced flight frequency with larger aircraft	
r. Multi-stage long distance travel	

16. What actions could the U.S. federal government undertake to promote the adoption of ...

a. Reduction of on-board weight	
b. Engine washing	
c. Limited use of paint on airframes	
d. Increased engine maintenance	
e. Single engine taxing	
f. Use of APU on ground at gate	
g. Automatic Dependent Surveillance – Broadcast (ADS-B)	
h. Required Navigation Performance (RNP)	
i. Continuous Descent Arrivals (CDA)	
j. NEXTGEN Weather	
k. Enhanced vision systems	
l. Synthetic vision systems	
m. Reduced Vertical Separation	
n. Flying at slower speeds	
o. Formation flying	
p. Air to air refueling	
q. Reduced flight frequency with larger aircraft	
r. Multi-stage long distance travel	

Part 3. Alternative Fuel Options

17. How would you rate your overall knowledge of alternative fuel options to reduce aircraft CO_2 emissions, such as biofuels?

1. None → SKIP TO QUESTION #25 7
2. Minimal → SKIP TO QUESTION #25
3. Basic → CONTINUE TO QUESTION #18
4. Proficient → CONTINUE TO QUESTION #18
5. Advanced → CONTINUE TO QUESTION #18

18. In your expert opinion, compared to jet fuel currently in use, what is the potential for future reduction of CO_2 emissions (on a life-cycle basis) for the following options?

	No potential	Low potential	Medium potential	High potential	Don't know
a. Coal to liquid	7	1	2	0	0
b. Fischer-Tropsch-treated feedstocks such as switchgrass	1	3	3	3	0
c. Fischer-Tropsch-treated forest waste	1	3	3	3	0
d. Fischer-Tropsch-treated municipal waste	1	5	1	3	0
e. Hydotreated algae	0	3	1	6	0
f. Hydrotreated Palm and Soy Oils	2	3	3	1	1
g. Hydrotreated camelina	1	2	3	2	2
h. Hydrotreated jatropha	0	3	3	2	2
i. Fuel cells	6	1	1	0	1
j. Hydrogen	4	2	1	1	1

19. In your expert opinion, what would be the potential R&D costs to develop the following options for commercial use?

	Low costs	Medium costs	High costs	Don't know
a. Coal to liquid	3	3	3	1
b. Fischer-Tropsch-treated feedstocks such as switchgrass	1	3	4	2
c. Fischer-Tropsch-treated forest waste	1	3	4	2
d. Fischer-Tropsch-treated municipal waste	0	4	4	2
e. Hydotreated algae	0	4	6	0
f. Hydrotreated Palm and Soy Oils	2	3	3	2
g. Hydrotreated camelina	2	3	2	3
h. Hydrotreated jatropha	2	3	2	3
i. Fuel cells	0	1	7	2
j. Hydrogen	0	1	8	1

20. In your expert opinion, what is the level of public acceptance for the following options?

	Low costs	Medium costs	High costs	Don't know
a. Coal to liquid	5	3	0	2
b. Fischer-Tropsc h-treated fe edstocks such as switchgrass	0	3	6	1
c. Fischer-Tropsch-treated forest waste	0	4	5	1
d. Fischer-Tropsch-treated municipal waste	0	3	6	1
e. Hydotreated algae	0	1	7	2
f. Hydrotreated Palm and Soy Oils	5	1	2	2
g. Hydrotreated camelina	0	5	2	3
h. Hydrotreated jatropha	0	3	4	3
i. Fuel cells	1	4	3	2
j. Hydrogen	5	2	0	3

21. In your expert opinion, given our best knowledge about future market conditions, and absent government intervention, how long would it take for the private sector to adopt these technologies?

	Never	Short timefame (<1 0 years)	Medium timeframe (10-20 years)	Long timeframe (> 20 years)	Don't know
a. Coal to liquid	7	2	0	0	1
b. Fischer-Tropsch-treated feedstocks such as switchgrass	5	4	0	0	1
c. Fischer-Tropsch-treated forest waste	5	4	0	0	1
d.Fischer-Tropsch-treated municipal waste	5	4	0	0	1
e. Hydotreated algae	3	4	2	0	1
f. Hydrotreated Palm and Soy Oils	5	3	0	0	2
g. Hydrotreated camelina	4	3	0	0	3
h. Hydrotreated jatropha	4	3	0	0	3
i. Fuel cells	0	3	6	0	1
j. Hydrogen	1	0	7	1	1

22. What major challenges exist to widespread use of ...

a. Coal to liquid	
b. Fischer-Tropsch-treated feedstocks such as switchgrass	
c. Fischer-Tropsch-treated forest waste	
d. Fischer-Tropsch-treated municipal waste	
e. Hydotreated algae	
f. Hydrotreated Palm and Soy Oils	
g. Hydrotreated camelina	
h. Hydrotreated jatropha	
i. Fuel cells	
j. Hydrogen	

23. What actions could the federal government undertake to promote the development and/or adoption of ...

a. Coal to liquid
b. Fischer-Tropsch-treated feedstocks such as switchgrass
c. Fischer-Tropsch-treated forest waste
d. Fischer-Tropsch-treated municipal waste
e. Hydotreated algae
f. Hydrotreated Palm and Soy Oils
g. Hydrotreated camelina
h. Hydrotreated jatropha
i. Fuel cells
j. Hydrogen

24. What other government actions, if any, should be undertaken to address greenhouse gas emissions from commercial aircraft ?

25. Do you have any other comments about anything covered in this rating tool? If so, please comment here.

APPENDIX IV: SCOPE AND METHODOLOGY

To address our objectives, we interviewed selected officials knowledgeable about the aviation industry, the industry's impact on the production of greenhouse gas and other emissions that have an impact on the climate, and options for reducing these emissions. We interviewed federal officials from the Environmental Protection Agency (EPA), FAA, the National Aeronautics and Space Administration (NASA) and the Departments of Defense and State. We also met with representatives of ICAO—a United Nations agency. We interviewed representatives of industry groups, environmental groups, airlines, aircraft manufacturers, aircraft engine manufacturers, alternative fuels manufacturers, economists, and academics. We interviewed officials based in the United States and abroad. We interviewed representatives of the EU and associations about the EU ETS. We completed a literature search and reviewed relevant documentation, studies, and articles related to our objectives. To specifically address commercial aviation's contribution to emissions, we asked our interviewees to identify the primary studies that estimate current and future emissions. As a result, we reviewed and summarized the findings of the 1999 International Panel of Climate Change Aviation and the Environment report and its 2007 Fourth Assessment Report, which were most frequently named as the most authoritative sources on global aviation emissions.

To specifically address technological and operational options to reduce commercial aviation's contribution to greenhouse gases and other emissions that can have an impact on the climate, we contracted with the National Academy of Sciences to identify and recruit experts in aviation and environmental issues. We interviewed 18 experts identified by the Academy, including those with expertise in aeronautics, air traffic management, atmospheric science, chemistry, climate change modeling, economics, environmental science, and transportation policy[180]. In conducting these interviews, we used a standardized interview guide to obtain consistent answers from our experts and had the interviews recorded and transcribed. Based on these interviews, we assembled a list of options for reducing aviation emissions, and we asked our experts to assess these options on several dimensions. We provided each of our experts with a standardized assessment tool that instructed the experts to assess the potential of each technological and operational option on the following dimensions: potential fuel savings and emissions reductions , potential research and development costs, potential cost to the airline industry, potential for public acceptance, and time frames for adoption. For each dimension, we asked the experts to assess each option on a three-point scale. For example, we asked the experts to rate each option as having "low potential", "medium potential", or "high potential" [181] for fuel savings and carbon dioxide emissions reductions.[182] We directed the experts not to answer questions about areas in which they did not have specific knowledge or expertise. As a result, throughout our report, the number of expert responses discussed for each emissions reduction option is smaller than 18, the number of experts we interviewed. Besides asking the experts to assess the potential of technological options, such as new aircraft and engine designs, we asked them to assess the potential of alternative fuels to reduce carbon dioxide emissions183. Furthermore, for operational options we asked the experts to assess included options that the federal government must implement, such as air traffic management improvements, as well as options that the airlines can exercise to reduce fuel burn. We analyzed and summarized the experts' responses in order to identify those technological and operational options that the experts collectively identified as holding the most promise for reducing emissions. To analyze the results, for each option and dimension, we counted the numbers of experts that selected the "low," " medium," and "high" responses. We then determined an overall, or group, answer for each question based on the response the experts most commonly selected for each option and dimension. However, if approximately the same number of experts selected a second response, then we chose both responses as the group answer. For example, rather than reporting that the experts rated a particular option as having "high" potential, we instead reported that they rated it as having "medium-high" potential if approximately the same number of experts selected the "high" response as selected the "medium" response. If approximately the same number experts selected all responses, we determined that there was no consensus on that question and reported the result as such[184].

In order to determine government options for reducing aviation emissions, we interviewed relevant experts, including those 18 recruited by the National Academy of Sciences, about the potential use and the costs and benefits of these options. We asked our interviewees to provide opinions and information on a variety of governmental options, including carbon taxes, cap-and-trade programs, aircraft and engine standards, government- sponsored research, and governmental subsidies. We looked at governmental actions that have been taken in the past and at those that have been proposed. We reviewed economic research on the economic impact of policy options for addressing greenhouse gas emissions. Our review focused on whether policy options could achieve emissions reductions from global sources in an economically efficient manner (for example, maximize net benefits). We interviewed EU officials to understand how the EU ETS will work and to determine issues related to this scheme, which is slated to include certain flights into and out of EU airports starting in 2012. Additionally, we reviewed and summarized the EU ETS and the legal implications of the scheme (see App. I).

APPENDIX V: COMMENTS FROM THE NATIONAL AERONAUTICS AND SPACE ADMINISTRATION

National Aeronautics and Space Administration
Headquarters
Washington, DC 20546-0001

May 26, 2009

Reply to Attn of: Aeronautics Research Mission Directorate

Ms. Cristina Chaplain
Director, Acquisition and Sourcing Management
United States Government Accountability Office
Washington, DC 20548

Dear Ms. Chaplain:

Thank you for the opportunity to review draft report, "Aviation and Climate Change: Aircraft Emissions Expected to Grow, But Technical and Operational Improvements and Government Policies Can Help Control Emissions," (GAO-09-554).

We found the report to be complete, concise, and accurate. In our opinion, it provides a balanced view of the issues related to the potential of future impact of aircraft emissions on climate change. Technical comments to the draft report have been provided separately.

Again, thank you for the opportunity to provide comments on the draft report and for your continued interest in aviation and its affect on climate change.

Sincerely,

Dr. Jaiwon Shin
Associate Administrator for
Aeronautics Research Mission Directorate

cc: Matthew Rosenberg

APPENDIX VI: COMMENTS FROM THE ENVIRONMENTAL PROTECTION AGENCY

UNITED STATES ENVIRONMENTAL PROTECTION AGENCY
WASHINGTON, D.C. 20460

MAY 2 1 2009

OFFICE OF
AIR AND RADIATION

Ms. Susan Fleming
Director, Physical Infrastructure Issues
U.S. Government Accountability Office (GAO)
441 G Street, N.W.
Washington, D.C. 20548

Dear Ms. Fleming:

Thank you for the opportunity to review and comment on the draft Government Accountability Office (GAO) report entitled, "Aviation and Climate Change: Aircraft Emissions Expected to Grow but Technological and Operational Improvements and Government Policies Can Help Control Emissions," which was provided to the U.S. Environmental Protection Agency (EPA) Administrator Lisa P. Jackson on May 1, 2009. We have reviewed and commented substantially on two earlier statements of facts related to this draft report.

While GAO has incorporated a portion of our written inputs on the initial and revised drafts, there are two major points which we still believe are not fully and correctly characterized within the report. The first is related to the potential role of emission standards as part of an overall response strategy and the second is related to the characterization of jet fuel consumption.

The Role of Emission Standards

The report fails to mention or discuss the extensive EPA Advanced Notice of Proposed Rulemaking (ANPRM) on greenhouse gas control under the Clean Air Act that was published in July 2008.[1] We strongly suggest this be referenced in the text. Further, we are concerned about the characterization of the potential role of emission standards in the underlined subtopic text on page 45 and the four paragraphs which follow. The report expresses the view that potential regulatory actions would limit technology responses and be economically inefficient. Based on our 35 years of experience in setting and implementing standards, we do not agree with this assessment. In fact, past EPA rules clearly show that flexible regulatory programs can be designed which optimize technology responses and yield benefits that far outweigh costs.[2]

[1] See the EPA Advance Notice of Proposed Rulemaking, "Regulating Greenhouse Gas Emissions Under the Clean Air Act", published in the Federal Register on July 30, 2008, 73 FR 44353.
[2] OMB has issued many Reports to Congress showing the economic benefits of EPA rules (benefits to society outweighing costs), for examples, see http://www.whitehouse.gov/omb/inforeg/regpol-reports_congress.html

Internet Address (URL) • http://www.epa.gov
Recycled/Recyclable • Printed with Vegetable Oil Based Inks on 100% Postconsumer, Process Chlorine Free Recycled Paper

Emission standards can be effectively used to limit emissions levels from aircraft. For mobile sources, EPA emission standards are typically not strict command and control requirements but are performance-based standards (not design or technology specific requirements) which provide manufacturers choices regarding which technology approaches to use to reduce emissions. In addition, emission standards typically include a variety of flexibility and incentive provisions to promote technology options and improve their overall economic efficiency by reducing costs.

These standards would include lead time and phase-in provisions (as in current aviation emission standards) to allow the manufacturers certainty and appropriate time to respond in the context of both market forces and future business needs. In addition, as is the case in most other EPA rules, they could include an emission averaging, banking, and trading program and other important flexibility provisions to optimize manufacturer technology response options, provide tools to improve compliance certainty, and reduce costs. Appropriate engine emission standards could drive measures for improved fuel specific fuel consumption and airframe-related standards could also reduce greenhouse gas emissions by improving efficiency. It now appears that the International Civil Aviation Organization will be considering both the engine and airframe approaches in the future and there is potential for EPA to do the same. With the projected increase in jet fuel consumption, achieving reductions in greenhouse gases will require consideration of technology measures and potentially market-based measures. Emission standards could in fact be complementary to the effective implementation of a market-based program.

Jet Fuel Consumption

Pages 4-7 discuss in detail the improved fuel efficiency of aviation operations. Aviation fuel efficiency has indeed improved significantly over the preceding three decades (see Figures 2 and 3 of the GAO report). However, there is not adequate discussion of the fact that the increase in aviation activity each year (more land-take-offs and more flying hours) results in an increase in total jet fuel consumption that has greatly outpaced the efficiency gains.

For example, according to data from the Air Transport Association, between 1978 and 2007, efficiency (as measured on a gallons per thousand revenue ton mile basis) improved by almost 110%.[3] However, according to information from the Bureau of Transportation Statistics, annual jet fuel consumption rose by 95% over that same period. As a result, jet fuel use was nine billion gallons higher in 2007 than in 1978.[4] Despite these noteworthy efficiency improvements, the annual Carbon Dioxide (CO2) inventory contribution from aviation was 94 million tons greater in 2007 than in 1978. Looking ahead, even with fleet turnover and the expected efficiency improvements from engine, air frame, and air traffic system upgrades in the future, the Federal Aviation Administration (FAA) projects an increase in jet fuel consumption of 31% (1.6% per year) between 2008 and 2025.[5] If these projections prove to be correct, the aviation jet

---

[3] See http://www.airlines.org/economics/energy/fuel+efficiency.htm, retrieved on May 15, 2009

[4] See http://www.transtats.bts.gov/fuel.asp?pn=1, retrieved on May 15, 2009

[5] See Table 22 from http://www.faa.gov/data_research/aviation/aerospace_forecasts/2009-2025/media/2009%20Forecast%20Doc.pdf

fuel consumption and carbon dioxide inventory contribution from aviation will be 31% greater in 2025 than in 2008, and 150% greater than in 1978.

Thank you for the opportunity to review and comment on this draft report. We look forward to our continued interactions with GAO and other stakeholders as we assess and consider policy options for addressing greenhouse gas emissions from aviation.

Sincerely,

Elizabeth Craig
Acting Assistant Administrator

END NOTES

[1] Kyoto Protocol to the United Nations Framework Convention on Climate Change (hereinafter the Kyoto Protocol). The Kyoto Protocol was adopted in December 1997 and was open for signature between Mar. 16, 1998, and Mar. 15, 1999. As of Jan. 14, 2009, 183 countries and the EEC had ratified the Kyoto Protocol. The binding emissions targets varies by country and is generally higher for more highly developed countries. For instance, the EEC has agreed to reduce their aggregate emissions by 8 percent from 1990 emissions levels.

[2] Kyoto Protocol, art. 2(2).

[3] This would also include those airports in non-EU countries currently participating in the EU ETS—Norway, Iceland, and Liechtenstein. See app. I for further explanation of the EU ETS.

[4] Fuel consumption of U.S.-based airlines roughly doubled during that same period.

[5] However, some aircraft available in the 1950s were about equally as fuel efficient as jets currently available today.

[6] According to IPCC, adaptation is an adjustment that occurs in response to expected or actual climatic stimuli or effects in order to moderate damages or exploit beneficial opportunities.

[7] IPCC defines mitigation as technological change and substitution that reduce resource inputs, such as energy use, and emissions per unit of output. Although several social, economic, and technological policies would produce an emissions reduction, with respect to climate change, mitigation means implementing policies to reduce greenhouse gas emissions.

[8] The PARTNER program—the Partnership for AiR Transportation Noise and Emissions Reduction—is a cooperative research organization sponsored by FAA, NASA, and Transport Canada, with members including universities and other organizations with expertise.

[9] NextGen is a new, satellite-based air traffic management system that is expected to increase the safety and enhance the capacity of the air transport system. NextGen will transform the current radar-based air traffic control system.

[10] 42 U.S.C. § 7571.

[11] However, EPA has the authority to set emissions standards. As the technical agency with responsibility for international civil aviation in the United States, FAA, in consultation with EPA, works with representatives of other ICAO member countries to formulate the standards.

[12] In addition, H.R. 2454, the American Clean Energy and Security Act of 2009, § 22 1(b), 111th Cong. (2009), would require EPA to issue standards for greenhouse gas emissions from new aircraft and new engines used in aircraft by Dec. 31, 2012.

[13] In addition, in 2007, the Supreme Court ruled that greenhouse gases meet the Clean Air Act's definition of an air pollutant and that EPA has the statutory authority to regulate greenhouse gas emissions from new motor vehicles under the Act. *Massachusetts v. Environmental Protection Agency*, 549 U.S. 497 (2007). As a result of the opinion, EPA must take one of three actions: 1) issue a finding that greenhouse gas emissions cause or contribute to air pollution that may endanger public heath or welfare; 2) issue a finding that greenhouse gases do not endanger public health or welfare; or 3) provide a reasonable explanation as to why it cannot or will not exercise its discretion to issue a finding. In response to this case, EPA issued an Advance Notice of Proposed Rulemaking, "Regulating Greenhouse Gas Emissions Under the Clean Air Act," 73 Fed. Reg. 44354 (July 30, 2008). EPA subsequently issued a proposed finding that carbon dioxide, methane, nitrous oxide, and hydrofluorocarbon emissions from new motor vehicles are contributing to air pollution which is endangering public health and welfare. 74 Fed. Reg. 18886 (Apr. 24, 2009). EPA is moving forward with the proposed finding while it develops proposed standards for regulating greenhouse gas emissions from motor vehicles.

[14] To date, 192 countries have ratified the UNFCCC.

[15] The emissions reduction goal varies by country and is generally higher for more highly developed countries. Some European countries, for example, agreed to reduce emissions by 8 percent compared to 1990 levels.

[16] In addition to the current 27 EU member states, other European countries not part of the EU, such as Norway, Iceland, and Liechtenstein, participate in the ETS. These airports are included in the amended ETS.

[17] Radiative forcing is a measure for the change of the Earth's energy balance due to a change in concentrations of greenhouse gases and other emissions that contribute to climate change.

[18] According to IPCC, the report was compiled by 107 lead authors from 18 countries. Successive drafts of the report were circulated for review by experts, followed by review of governments and experts. Over 100 contributing authors submitted draft text and information to the lead authors and over 150 reviewers submitted suggestions for improvement during the review process.

[19] According to IPCC, global aviation's global carbon dioxide emissions totaled an estimated 480 million tons in 2000.

[20] According to EPA, in 2007 commercial aviation represented about 82 percent of all aviation emissions, and about 2.5 percent of U.S. carbon dioxide emissions. In the United States, military aviation represents less than 10 percent of total aviation emissions.

[21] EPA, *Draft Inventory of U.S. Greenhouse Gas Emissions and Sinks: 1990-2007*, (March 2009). The aviation estimates do not include fuel consumed for international air transport per UNFCCC's reporting guidelines; according to the EPA, in 2006, when those international fuels were included, domestic and international commercial, military, and general aviation flights represented about 3.4 percent of U.S. carbon dioxide emissions.

[22] In the United States according to EPA, carbon dioxide emissions from fossil fuel combustion in the transportation sector account for an estimated 31 percent of carbon dioxide emissions (in carbon dioxide equivalents not including fuels used for international flights). Carbon dioxide equivalents provide a common standard for measuring the warming potential of different greenhouse gases and are calculated by multiplying the emissions of the non-carbon-dioxide gas by its global warming potential, a factor that measures its heat-trapping ability relative to that of carbon dioxide.

[23] This 3 percent radiative forcing estimate for aviation is based on more recent research, included in IPCC's 2007 *Fourth Assessment Report*. Originally, IPCC calculated a 3.5 percent estimate for 1992, which it published in its 1999 special report, *Aviation and the Global Atmosphere*.

[24] See *Aviation Climate Change Research Initiative: A Report on the Way Forward,* (2008) citing Forster, P., et al. (2007), Changes in atmospheric constituents and in radiative forcing. In: IPCC, 2007: Climate Change 2007: The Physical Science Basis. Contribution of Working Group I to the Fourth Assessment Report of the Intergovernmental Panel on Climate Change [Solomon, S., D. Qin, M. Manning, Z. Chen, M. Marquis, K.B. Averyt, M. Tignor and H.L. Miller (eds.)]. Cambridge University Press, Cambridge, United Kingdom and New York, NY, USA; and Sausen, R. et al., *Aviation radiative forcing in 2000: An update on IPCC* (1999), Meteorologische Zeitschrift, 14-4, 555–561, 2005.

[25] The 1999 IPCC report originally estimated that aviation's impact could be two to four times that of carbon dioxide alone. The use of a radiative forcing ratio of the forcing of all aviation emissions over carbon dioxide is an approximation and is an area for further research and understanding.

[26] IPCC assumed a maturing global market with the above economic growth rates from 1990 to 2025 and slightly lower economic growth rates thereafter.

[27] This 5 percent estimate of aviation's contribution to 2050 global radiative forcing is based on IPCC's midrange estimate. IPCC did not provide this estimate for the 2050 high- and low-range scenarios.

[28] B. Owen and D.S. Lee, *Allocation of International Aviation Emissions from Scheduled Air Traffic—Future Cases, 2005 to 2050*, Manchester Metropolitan University (Manchester, UK, March 2006) and Ralf Berghof, Alf Schmitt et al., *CONSAVE 2050 Final Technical Report* (July 2005).

[29] The 1999 IPCC report included a high-growth emissions scenario for 2050 from an environmental organization that showed aviation contributing 9.8 percent of global carbon dioxide emissions. This estimate assumed air traffic levels for 2050 that were more than double those used in the highest IPCC estimate. IPCC concluded that such an estimate for 2050, though not impossible, was unlikely because it would require countries to build more than 1,300 new airports with 15 gates each, or the equivalent of 2 new airports per month for 60 years, which would be unprecedented compared with historical increases in global airport capacity.

[30] This estimate is based on emissions of carbon dioxide and nitrogen oxides. In terms of fuel consumption, FAA estimates a 28 percent increase in US carriers commercial aviation jet fuel from 2001 to 2025 (see *FAA Aerospace Forecast, Fiscal Years 2009-2025*, table 22).

[31] Two main methods, market exchange rates and purchasing power parity are used to convert the GDP of a country in national currency terms to a common currency, usually the U.S. dollar. GDP growth rates can differ depending on the conversion method used.

[32] However, the forecast doesn't account for the possibility that some airlines might adopt low-carbon alternative fuels.

[33] The United States is currently working on its future air traffic management system, NextGen, as is Europe through the Single European Sky Air Traffic Management Research Program (SESAR).

[34] Other studies have estimated aviation's share of current climate change emissions and find aviation's contributions to climate change to be larger when the impact of cirrus clouds is included. For example, one environmental organization in a 2006 report included the upper range for aviation-induced cirrus clouds' radiative forcing (from Sausen et al, 2005, which was reported in IPCC) in its aviation radiative forcing total and calculated that aviation contributed 9 percent of total radiative forcing due to human activity worldwide in 2000.

[35] Joosung J. Lee, Stephen P. Lukachko, Ian A. Waitz, and Andreas Schafer, "Historical and Future Trends in Aircraft Performance, Cost, and Emissions," *Annual Reviews on Energy and Environment*, vol. 26 (2001).

[36] The higher the bypass ratio, the more air bypasses the engine, providing extra propulsion for the engine fan blades and increasing the engine's efficiency.

[37] In addition, General Electric is developing a new fuel-efficient engine, the GEnx. According to General Electric, this engine will improve fuel burn by 15 percent, while also reducing NOx emissions, compared with General Electric's previous generation of engine technologies.

[38] The U.S. Air Force also is conducting research into more fuel efficient engines through its ADVENT program, a joint effort with engine manufacturers that Air Force officials told us is expected to have commercial applications.

[39] However, National Aeronautics and Space Administration staff told us that no such investments will be necessary.

[40] For more information on challenges in NextGen implementation see GAO, *Next Generation Air Transportation System: Issues Associated with Midterm Implementation of Capabilities and Full System Transformation*, GAO-09-481T (Washington, D.C.: Mar. 25, 2009).

[41] Although plants converted into biofuels remove carbon dioxide from the air during growth, the biofuel production process is not emissions-free. For instance, nitrous oxide— a powerful greenhouse gas—may be emitted when nitrogen-based fertilizers are applied to soils to increase yields of important biofuel feedstocks such as corn. In addition, fossil fuels are burned during the harvesting, transporting and refining of biofuel feedstock plants. Finally, researchers have raised concerns that increased biofuel production could result in additional greenhouse gas impacts due to the conversion of lands not previously used for biofuel crop production.

[42] On May 26, 2009, EPA issued a Notice of Proposed Rulemaking regarding the renewable fuel standard, as required by the Energy Independence and Security Act of 2007 (EISA). When finalized, the proposed rule would implement EISA's changes to the renewable fuel standard. 74 Fed. Reg. 24904 (May 26, 2009).

[43] In addition, FAA and the U.S. Air Force, through the Partnership for AiR Transportation Noise and Emissions Reduction (PARTNER) program—an FAA-NASA-Transport Canada-sponsored Center of Excellence, is sponsoring a project that aims to develop a tool that can estimate the life-cycle environmental impact from alternative jet fuels. The U.S. Air Force is leading an interagency working group including EPA, FAA, the Department of Energy and researchers to develop a document of life-cycle analysis best practices.

[44] While in 2008 Boeing conducted a test flight of a small aircraft powered by hydrogen fuel cells, the manufacturer indicated it does not expect that large aircraft will be able to derive their primary energy from such sources.

[45] For a discussion of biofuel production see GAO, *Biofuels: DOE Lacks a Strategic Approach to Coordinate Increasing Production with Infrastructure Development and Vehicle Needs* GAO-07-713 (Washington, D.C.: June 8, 2007).

[46] However, according to DOT, it could also be argued that because aircraft do not face the potential for non-liquid fuel sources of energy—for example, electricity—and because there is a proportionally higher premium for refining jet fuel, aviation may be a more certain consumer of biofuel supplies.

[47] Algae-based biofuel composed only 2.5 percent of the fuel blend.

[48] This analysis assumed that biofuels are first available in 2015 and their usage increases at a rate of 2 percent per year, and through life-cycle carbon emissions, biofuels emit 20 percent the carbon dioxide as jet fuel. For blended-wing body aircraft, the analysis assumed an 18 percent fuel burn reduction and introduction of the aircraft into global fleets in 2025 with a 30-year time frame to achieve a penetration of 33 percent of global aircraft fleets. The analysis also assumed the adoption of other improvements, including the use of lightweight materials, open rotor engines, formation flying, and multistage long- distance travel. We did not review the reliability of this analysis.

[49] A lower assumed traffic growth rate would have reduced the study's forecast of emissions. With respect to biofuels, this study assumed that biofuels would be used between 2010 and 2050 (at which time they would represent 25 percent of aviation fuels) and would produce

10 percent of the carbon dioxide emissions of fossil fuels. We did not review the reliability of this analysis.

[50] For a comprehensive discussion of the economics of policy options to address climate change as well as discussion of a combination of such options, see GAO, *Climate Change: Expert Opinion on the Economics of Policy Options to Address Climate Change*, GAO-08-605 (Washington, D.C.: May 9, 2008).

[51] However, to the extent that a cap-and-trade program regulates other transportation sectors as well, costs of using alternative transportation modes are likely to be higher as well.

[52] For example, the demand for energy may increase during unexpectedly hot or cold periods, leading to price spikes and making it more expensive for sources to meet the cap.

[53] The safety valve would help prevent the price of allowances from exceeding the expected benefits of the emissions reductions. Under a cap-and-trade program, the emissions cap could be set at a level that balances the expected marginal cost of meeting the cap with an estimate of the marginal benefits. The safety valve price could be set just above the expected marginal costs to avoid the possibility that the cap is overly stringent.

[54] We have reported on the importance of setting an emissions baseline. For example, setting a baseline with poor historical data could lead to the creation of a baseline that is above actual emissions, leading to no emission reductions. See GAO, *Climate Change Science: High Quality Greenhouse Gas Emissions Data Are a Cornerstone of Programs to Address Climate Change*, GAO-09-423T (Washington, D.C.: Feb. 24, 2009).

[55] However, some interest groups have expressed concern that an emissions cap does not take into account emissions reductions that airlines have achieved in recent years and claim that it unfairly penalizes those airlines that have reduced emissions.

[56] For example, Association of European Airlines expressed a concern that because some countries in Europe have taxes on aircraft carbon dioxide emissions, airlines may be subject to those taxes as well as the ETS, meaning that their emissions would be counted twice by two different regulations.

[57] We have reported on lessons learned from the EU ETS. See GAO, *International Climate Change Programs: Lessons Learned from the European Union's Emissions Trading Scheme and the Kyoto Protocol's Clean Development Mechanism,* GAO-09-151 (Washington, D.C.: Nov. 18, 2008).

[58] With respect to aircraft, the ETS currently regulates only carbon dioxide emissions. However, according to the European Commission, it has agreed to address nitrogen oxide emissions from aircraft through a separate legislative measure. As of May 2009, no EU legislative proposal has specifically addressed nitrogen oxide emissions from aircraft.

[59] For example, Japan is considering creating a voluntary cap-and-trade program and Japan Airlines already has indicated that it will participate in the scheme, but only based on its domestic flights.

[60] H.R. 1862, 111th Cong. (2009).

[61] S. 3036, 110th Cong. (2008).

[62] For example, the social cost of carbon reflects the present value of economic damages caused by an additional quantity of emissions. Under an economically optimal policy, the price would be set at a point where the marginal damages from global warming equal the marginal cost of controlling emissions.

[63] A metric ton equals 2,205 pounds.

[64] The extent to which a cap-and-trade program would provide greater certainty about emissions reductions depends on the design of the particular program. For example, some programs may include cost containment measures that allow total emissions to exceed the cap under certain conditions.

[65] Price elasticity of demand is a measure of the responsiveness in quantity demanded as a result of a change in price. Goods and services with a high elasticity of demand will see larger changes in the quantity demanded than the change in price.

[66] Equivalent to about $4.5 billion based on exchange rates on Apr. 20, 2009.

[67] According to industry interests groups we met with, to the extent that airlines incur the costs of a tax or for cap-and-trade allowances, their resources for upgrading their fleets with more fuel-efficient aircraft and for implementing other emissions-reduction measures will be more limited.

[68] Equivalent to about $11.60 based on exchange rates on April 20, 2009.

[69] Some organizations we met believe that revenues collected through the auctions of allowances should be used for climate change mitigation and adaptation as well as to help fund low-emissions technologies.

[70] H.R. 2454, § 22 1(b), 111th Cong. (2009).

[71] Emissions standards typically do not equalize marginal costs across sources, a basic condition for efficiency. In the case of greenhouse gas emissions, emissions are uniformly mixed and abatement costs vary widely across sources. These characteristics favor market- based instruments, which can achieve significant cost savings by encouraging low-cost sources to make the bulk of the emissions reductions.

[72] According to FAA and EPA, existing aircraft engine standards do not mandate specific technologies and are developed on the basis of technological practicability.

[73] According to EPA, well designed standards could give aircraft engine manufacturers the flexibility to make cost-effective reductions. In addition, according to an official in EPA's Office of Air and Radiation, with the projected increase in jet fuel consumption and related emissions, addressing greenhouse gas emissions will require consideration of technology measures and market-based measures. However, under a market-based program such as cap-and-trade, emissions caps set the total emissions level and sources would need to determine how to accommodate business growth while complying with the cap. Such an approach gives sources an incentive to innovate and search for low-cost ways to reduce emissions.

[74] In addition, the U.S. Air Force conducts research that has potential for applicability for commercial aviation. For example, through the ADVENT program, the Air Force is researching fuel-efficient engine technologies.

[75] NASA refers to the three generations of technologies as N+1, N+2, and N+3.

[76] The plan also sets goals for future noise levels and nitrogen oxide emissions.

[77] National Science and Technology Council, *National Plan for Aeronautics Research and Development and Related Infrastructure* (Washington, D.C., Jan. 18, 2008).

[78] GAO, *Aviation and the Environment: FAA's and NASA's Research and Development Plans for Noise Reduction Are Aligned, but the Prospects of Achieving Noise Reduction Goals Are Uncertain*, GAO-08-384 (Washington, D.C., Feb. 15, 2008).

[79] GAO, *Aviation and the Environment: NextGen and Research and Development Are Keys to Reducing Emissions and Their Impact on Health and Climate*, GAO-08-706T (Washington, D.C., May 6, 2008).

[80] About $200 million to $267 million per year based on exchange rates on Apr. 29, 2009.

[81] Equivalent to approximately $2.1 billion based on exchange rates on Mar. 31, 2009.

[82] The European Commission initiates the legislative process by drafting specific pieces of legislation and proposing them to the Council of the European Union and European Parliament, who together serve as the EU's legislative branch.

[83] Under the EU's co-decision procedure, the legislative procedure for environmental and certain other types of laws, both the Council of the European Union and European Parliament must approve legislation in order to enact a law. Once both bodies approve identical texts of the legislation, it must be published in the Official Journal of the European Union. The law goes into force 20 days after publication. Thereafter, each member state has 1 year to transpose the directive into national law.

[84] Directive 2008/101/EC, 2009 O.J. (L 8) 3. The directive was amended on Mar. 26, 2009.

[85] Some flights are excluded from the cap, including military flights, flight operations for emergency purposes, such as firefighting, as well as airlines with very limited operations. *See* 2008/101/EC, 2009 O.J. (L 8) 3, Annex I.

[86] For instance, a flight from Los Angeles to London will have to surrender allowances for its travel in U.S. airspace, international airspace, and U.K. airspace.

[87] Historical aviation emissions is defined as the mean average of the annual emissions in calendar years 2004, 2005, and 2006. This will be the baseline for emissions reductions. *See* Directive 2008/101/EC, 2009 O.J. (L8) 3, art. 3c.

[88] *See* Directive 2008/101/EC, art. 3(c) (establishing the cap). The cap for subsequent trading periods can be adjusted by an amendment to the directive. *See also* Proposal for a Directive COM(2008) 16 (extending the trading periods to 8 years from 5 years). Although both the European Parliament and the European Council have signaled final approval of the proposed directive, final, formal approval had not occurred as of June 2, 2009.

[89] Three percent of the total quantity of allowances to be allocated will be set aside in a special reserve for new aircraft operators or aircraft operators with rapid growth. Eligible aircraft operators must apply to their assigned member states to obtain free allowances from the special reserve. The European Commission will determine how the allowances in the special reserve will be distributed. Any unallocated allowances will be auctioned.

[90] Aircraft operators will be assigned to the member state that either: (1) issued its operating license or (2) has the greatest estimated emissions from flights performed by that aircraft operator in 2006 or the operator's first year of operation. *See* Directive 2008/101/EC, 2009 O.J. (L 8) 3, art. 18(a). These assignments were made in February 2009 and most U.S. airlines were assigned to the United Kingdom. According to the Directorate-General of the Environment (DG Environment), the auctions will be open to anyone to participate and the Commission is currently developing the regulation containing detailed provisions for member state auctions.

[91] For 2012, each member state's total allowances will be based on verified 2010 emissions. In subsequent years, the total allowances will be based on verified emissions from 2 years prior.

[92] According to the 2008 directive, it shall be for member states to determine the use to be made of revenues generated from the auctioning of allowances. Those revenues should be used to tackle climate change in the EU and third countries, inter alia; to reduce greenhouse gas emissions; to adapt to the impacts of climate change in the EU and third countries, especially developing countries; to fund research and development for mitigation and adaptation, including in particular in the fields of aeronautics and air transport; to reduce emissions through low-emission transport; and to cover the cost of administering the Community scheme.

[93] Alternatively, if an aircraft operator has excess emissions allowances, it will be able to sell those excess allowances on the market.

[94] The European Commission will determine the exact percentage of CERs and ERUs that aircraft operators can use in the 2013 through 2020 trading period. In determining the percentage of CERs and ERUs that each covered sector can use in the 2012 through 2020 trading period, the Commission will ensure that overall CER and ERU usage does not exceed 50 percent of the emissions reductions achieved in Phase II (2007 through 2012), measured from a baseline of 2005 emissions levels.

[95] According to the DG Environment, draft legislation and a number of bills proposed in the 110th Congress would have triggered such consideration. *See* McCain-Lieberman, S. 280, 110th Cong. (2007); Lieberman-Warner, S. 2191, 110th Cong. (2007); Dingell-Boucher, draft bill; and Waxman-Markey, H.R. 1590, 110th Cong. (2007), H.R. 6186, 110th Cong. (2008).

[96] On Feb. 11, 2009, the Commission adopted the Preliminary List of Aircraft Operators and their administering Member States, *Commission Notice Pursuant to Article 18a(3)(a) of Directive 2003/87/EC*, C(2009) 866. Additional preparations have included holding a number of stakeholder workshops to discuss implementing this directive and development of guidelines for monitoring, reporting, and verification of emissions, according to the DG Environment.

[97] All operators must submit their monitoring plan by the end of August 2009, according to the DG Environment.

[98] United Nations Framework Convention on Climate Change, 1992, Article 2, GE.05-62220. (E) 200705.

[99] *Id.* at Article 4(1)(b).

[100] *Id.* at Article 4.

[101] Kyoto Protocol to the United Nations Framework Convention on Climate Change, 1999 [hereinafter Kyoto Protocol].

[102] The Kyoto Protocol recognized that industrialized nations are the largest contributors to greenhouse gas emissions and therefore divided the signatory countries into two groups—Annex I countries that are industrialized nations subject to binding targets, either emission reduction or limitation requirements, and Non-Annex I countries that are not subject to binding targets.

[103] Kyoto Protocol, *supra* note 20, art. 3(1).

[104] *Id.* at art. 2(2).

[105] Convention on International Civil Aviation, Dec. 7, 1944, Ninth ed., 2006, 61 Stat. 1180, 15 U.N.T.S. 295 [hereinafter Chicago Convention].

[106] *See Id.* at art. 1.

[107] Article 11 provides that "[S]ubject to the provisions of this Convention, the laws and regulations of a contracting State relating to the admission to or departure from its territory of aircraft engaged in international navigation . . . shall be applied to the aircraft of all contracting States without distinction as to nationality, and shall be complied with by such aircraft upon entering or departing from or while within the territory of that State."

[108] Article 12 states that "[e]ach contracting State undertakes to adopt measures to insure that every aircraft flying over or maneuvering within its territory and that every aircraft carrying its nationality mark, wherever such aircraft may be, shall comply with the rules and regulations relating to the flight and maneuver of aircraft there in force. . . . Over the high seas, the rules in force shall be those established under this Convention, . . ."

[109] Article 15 provides in part that "[n]o fees, dues or other charges shall be imposed by any contracting State in respect solely of the right of transit over or entry into or exit from its territory of any aircraft of a contracting State or persons or property thereon." Moreover, charges that may be imposed must be non-discriminatory.

[110] Article 24 covers customs and related duties for aircraft of contracting states and includes the provision that "[f]uel . . . on board an aircraft of a contracting State, on arrival in the territory of another contracting State and retained on board on leaving the territory of that State shall be exempt from customs duty, inspection fees or similar national or local duties and charges."

[111] Consolidated Statement on Continuing ICAO Policies and Practices Related to Environmental Protection, ICAO Res. A36-22, 36th Session, Appendix L, 1(b)(1)(2007) [hereinafter ICAO Res. A36-22].

[112] Many positions referred to in this article cite to ICAO Resolution A35-5 rather than the A36-22.

[113] ICAO Res. A36-22, *supra* note 30, Appendix K.

[114] This also is referred to as the U.S.-EU Open Skies Agreement.

[115] *See generally* Id.

[116] *Id.* at art. 12, 3.

[117] Bali Action Plan, Decision 1/CP.13, 1 (FCCC/CP/2007/6/Add.1).

[118] Letter of U.S. Ambassador Kristen Silverberg to Jos Delbeke, Acting Director General General, Environment Directorate General, Oct. 30, 2008; and letter from the Ambassadors of Australia, Canada, China, Japan, Korea and the United States to Ambassador Peter Witt of Germany, Apr. 6, 2007 [hereinafter Ambassador Letter].

[119] *See* statement of Secretary Clinton: "President Obama and I recognize that the solutions to this crisis are both domestic and global, that all nations bear responsibility and all nations must work together to find solutions." *See also* statement of Special Envoy Stern: "Yet we can only meet the climate challenge with a response that is genuinely global. Eighty percent of greenhouse gas emissions are produced outside the United States, and a rapidly growing percentage is produced in emerging market countries."

[120] ATA's analysis is contained in a non-public Legal Analysis Summary provided to GAO.

[121] Chicago Convention, *supra* note 24, art. 11.

[122] *Id.* at art. 15.

[123] Article 24 provides that "fuel...on board an aircraft of a contracting State, on arrival in the territory of another... shall be exempt from customs duty, inspection fees or similar national or local duties and charges."

[124] Chicago Convention, *supra* note 24, art. 1. Article 1 provides that every contracting state has "complete and exclusive sovereignty over the airspace above its territory."

[125] *Id.* at art. 12.

[126] *Id.* at art. 15.

[127] *See* Josh Voorhees, Greenwire, *EU brings airlines into climate scheme; industry vows court fight* (2008), at http://www.eenews.net/Greenwire/2008/07/08/archive/3.

[128] Press Release, Air Transport Association Chief Executive Says European Aviation Emissions Trading Scheme 'Contrary to International Law and Bad Policy' (Oct. 30, 2008), *available at* http://www.airlines.org/news/releases/2008/news_10-30-08.htm.

[129] A "Sense of the Congress" is not legally binding if passed.

[130] H.R. 915, Sec. 514(1), 111th Cong. (2009).

[131] *Id.* at Sec. 514(2).

[132] The legal assessment was conducted by an independent contractor for the DG Environment.

[133] IETA, *IETA's position on the Inclusion of Aviation in the EU ETS*, available at www.ieta.org/ieta/www/pages/getfile.php?docID=2413.

[134] *Id.* at 2.

[135] *Id.*

[136] The legal assessment was conducted by an independent contractor for the DG Environment.

[137] *Giving Wings to Emissions Trading, Inclusion of Aviation under the European emissions trading system (ETS): Design and Impacts* (The Delft Report), Report to the European Commission, DG Environment, No.ENV.C.2./ETU/2004/0074r, July 2005 [hereinafter Delft Report]. This report was prepared for and endorsed by the European Commission.

[138] *Id.* at 170-73.

[139] Chicago Convention, *supra* note 24, art. 11.

[140] Article 12 requires contracting states to ensure that aircraft under their jurisdiction are in compliance with rules and regulations relating to the flight and maneuver of aircraft.

[141] Chicago Convention, *supra* note 24, art. 15.

[142] *Id.* at art. 24.

[143] The Delft Report at 175-77.

[144] *Id.* at 176-77

[145] *Id.* at 177.

[146] Consolidated statement of continuing ICAO policies and practices related to environmental protection, ICAO Res. A35-5, 35th Session, Appendix 1 (2004) [hereinafter ICAO Res. A35-5]. "ICAO endorses an open emission system for international aviation" and requests council focus on two areas under this framework.

[147] The Delft Report further elaborates: "the operation of an ETS does not unilaterally limit the volume of traffic ... or aircraft types, as it only provides incentives to reduce emissions over time; fair and equal opportunity to participate in the ETS is already covered under the Chicago Convention's non-discrimination clause and this must be adhered to under the ETS." *See* pp. 179-181.
[148] Letter to the German Presidency regarding Commission proposal to extend ETS to aviation (Apr. 6, 2007).
[149] *Id.*
[150] *Id.*
[151] *Id.*
[152] *Id.*
[153] For brevity's sake, we have not included explanatory articles because they are repetitive of what is contained in this appendix. Here we highlight those articles which set forth legal positions on the issue.
[154] Daniel B. Reagan, *Putting International Aviation Into the European Union Emissions Trading Scheme: Can Europe Do It Flying Solo?,* 35 B.C. Envtl. Aff. L. Rev. 349 (2008).
[155] *Id.* at 380-8 1.
[156] *Id.* at 38 1-82.
[157] *Id.* at 382.
[158] *Id.* at 383.
[159] Gisbert Schwarze, *Including Aviation into the European Union Emissions Trading Scheme*, European Environmental Law Review (2007).
[160] *Id.* at 12-15.
[161] *Id.* at 12-13.
[162] *Id.* at 13-14.
[163] In the alternative, as argued, even if this did constitute a charge, Article 15's only applicability is to discrimination as to nationality. *Id.*
[164] Gisbert Schwarze, *Including Aviation into the European Union Emissions Trading Scheme*, p. 13-14, European Environmental Law Review (2007).
[165] *Id.* at 14.
[166] *Id.*
[167] *Id.*
[168] *Id.*
[169] United Nations Framework Convention on Climate Change, *supra* note 17, art. 4(2)(a).
[170] Kyoto Protocol, *supra* note 20, art. 2(2).
[171] Schwarze at 13. *See also* ICAO Res. A35-5, *supra* note 66, Appendix I Nr. 2(c)(1).
[172] *See* Voorhees, *supra* note 46.
[173] *See* Ambassadors Letter, *supra* note 37.
[174] Chicago Convention, *supra* note 24, art. 84. The Council is composed of 36 contracting states elected by the Assembly. *See Id.* at art. 50. When a disagreement is submitted to the Council, the Council will invoke the Rules of the Settlement of Differences, established in 1957, which sets forth procedures for disputes.
[175] *Id.* at art. 84. Additional requirements for settlement of disputes and penalties are contained in Articles 85, 86, 87 and 88 of the Chicago Convention. Alternatively, a challenge could potentially be brought against a carrier. If a carrier chooses to move forward and not comply with the ETS, action could be brought under the Chicago Convention Article 87 and force the airline to cease operations over any contracting state.
[176] *See* U.S.-EU Air Transport Agreement, art. 19.
[177] *See Id.* at art. 19(7).
[178] *See Aimee Turner, Flight Global, EU warns ICAO it will go it alone if no progress is made with ETS* (2007), available *at* www.flightglobal.com.
[179] Id.

[180] See Appendix II for a list of the experts we interviewed.

[181] We did not provide specific definitions for "low," "medium," and "high" and let each respondent determine what it meant to them.

[182] See app. III for a copy of the assessment tool that we asked the experts to complete as well as complete results. Additionally, we asked the experts to describe the major challenges to the widespread use of the options and the actions that the U.S. federal government could take to promote the development and/or adoption of any of these options.

[183] For alternative fuel options, we used a four-point scale to assess each option's potential to reduce carbon dioxide emissions ("no potential," "low potential," "medium potential "or "high potential").

[184] We defined "approximately the same number of experts" as being either the exact same number of experts or as one fewer or one more expert.

In: Aviation and Climate Change
Editors: George T. Blumenthal
ISBN: 978-1-60876-757-1

Chapter 2

AVIATION & EMISSIONS – A PRIMER

Federal Aviation Administration

Aviation plays a key role in the economic prosperity and lifestyle Americans enjoy. Our economy benefits greatly from the ability to move people and products all over the globe - quickly and safely. Aviation contributes to our quality of life - allowing us to visit friends and relatives, to travel, to experience new places, to shrink the borders of the world. The statistics are impressive. In 1903 the year of the Wright brothers' first flight, earth's population was 1.6 billion[1]; today, over 1.6 billion people use the world's airlines[2]. The air transport industry provides 28 million direct, indirect, and induced jobs worldwide[3]. And aircraft carry about 40% of the value of all world trade[4], driving the "just in time" deliveries critical to productivity improvements.

Air transport links our world and is a key tenet of continued economic development and security for the U.S. However, aviation also has environmental impacts – primarily noise and atmospheric emissions. While aircraft noise issues are better known, less focus has been placed on emissions. This paper provides a brief overview of important issues regarding aviation emissions.

WHAT EMISSIONS COME FROM AVIATION?

Aircraft produce the same types of emissions as your automobile. Aircraft jet engines, like many other vehicle engines, produce carbon dioxide (CO_2), water vapor (H_2O), nitrogen oxides (NO_N), carbon monoxide (CO), oxides of sulfur (SO_N), unburned or partially combusted hydrocarbons (also known as volatile organic compounds (VOCs)), particulates, and other trace compounds. A small subset of the VOCs and particulates are considered hazardous air pollutants (HAPs). Aircraft engine emissions are roughly composed of about 70 percent CO_2, a little less than 30 percent H_2O, and less than 1 percent each of NO_N, CO, SO_N, VOC, particulates, and other trace components including HAPs. Aircraft emissions, depending on whether they occur near the ground or at altitude, are primarily considered local air quality pollutants or greenhouse gases, respectively. Water in the aircraft exhaust at altitude may have a greenhouse effect, and occasionally this water produces contrails, which also may have a greenhouse effect. About 10 percent of aircraft emissions of all types, except hydrocarbons and CO, are produced during airport ground level operations and during landing and takeoff. The bulk of aircraft emissions (90 percent) occur at higher altitudes. For hydrocarbons and CO, the split is closer to 30 percent ground level emissions and 70 percent at higher altitudes.

Aircraft are not the only source of aviation emissions. Airport access and ground support vehicles produce similar emissions. Such vehicles include traffic to and from the airport, ground equipment that services aircraft, and shuttle buses and vans serving passengers. Other emissions sources at the airport include auxiliary power units providing electricity and air conditioning to aircraft parked at airport terminal gates, stationary airport power sources, and construction equipment operating on the airport.

WHAT DETERMINES AVIATION EMISSIONS?

Aviation emissions reflect the level of overall aviation activity. The growth of air travel for the past several decades has been very rapid. Demand for travel services, both passenger travel and freight transportation, is increasing substantially. According to the U.S. Bureau of Transportation Statistics[5] a 21.5 percent increase in population, 32 percent increase in the labor force, and 90 percent increase in GDP between 1980 and 2000 have driven this demand. The chart[6] below shows the growth of aviation and the economy. Demand for air travel grows as the economy grows and prosperity increases.

EMISSIONS FROM COMBUSTION PROCESSES

CO_2 – Carbon dioxide is the product of complete combustion of hydrocarbon fuels like gasoline, jet fuel, and diesel. Carbon in fuel combines with oxygen in the air to produce CO_2 .

H_2O – Water vapor is the other product of complete combustion as hydrogen in the fuel combines with oxygen in the air to produce H_2O.

NO_x – Nitrogen oxides are produced when air passes through high temperature/high pressure combustion and nitrogen and oxygen present in the air combine to form NO_x.

HC – Hydrocarbons are emitted due to incomplete fuel combustion. They are also referred to as volatile organic compounds (VOCs). Many VOCs are also hazardous air pollutants.

CO – Carbon monoxide is formed due to the incomplete combustion of the carbon in the fuel.

SO_x – Sulfur oxides are produced when small quantities of sulfur, present in essentially all hydrocarbon fuels, combine with oxygen from the air during combustion.

Particulates – Small particles that form as a result of incomplete combustion, and are small enough to be inhaled, are referred to as particulates. Particulates can be solid or liquid.

Ozone – O_3 is not emitted directly into the air but is formed by the reaction of VOCs and NO_x in the presence of heat and sunlight. Ozone forms readily in the atmosphere and is the primary constituent of smog. For this reason it is an important consideration in the environmental impact of aviation.

Over the long term, we expect that demand for air transportation will continue to grow rapidly to support our economic productivity, our quality of life, and our national security. More and more the worldwide transportation system is becoming an integrated transportation network. For most long-distance travel, however, aviation's speed, convenience, and cost overcome consideration of other travel modes. It only faces competition on short trips or when moving low value or high volume products. Looking to the future, the forecast is for continued strong growth as shown in the following graph[7]. This is consistent with the demand for transportation generally, which is increasing largely in response to very positive structural changes in both the domestic and global economies.

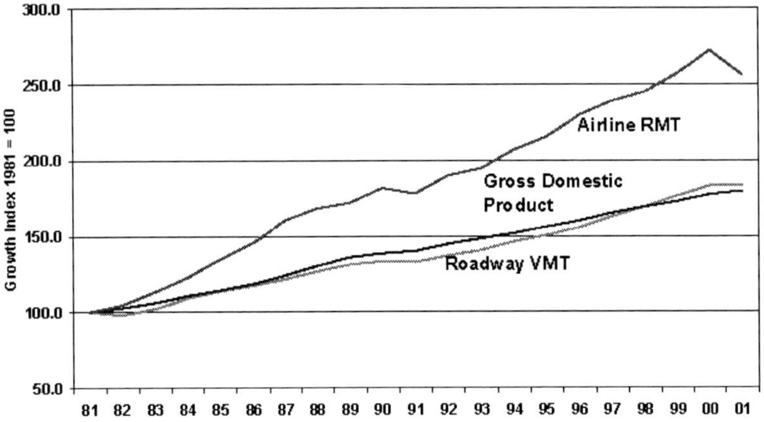

For 20 years aviation growth in the U.S. outpaced the growth of the economy although all transportation has grown significantly.

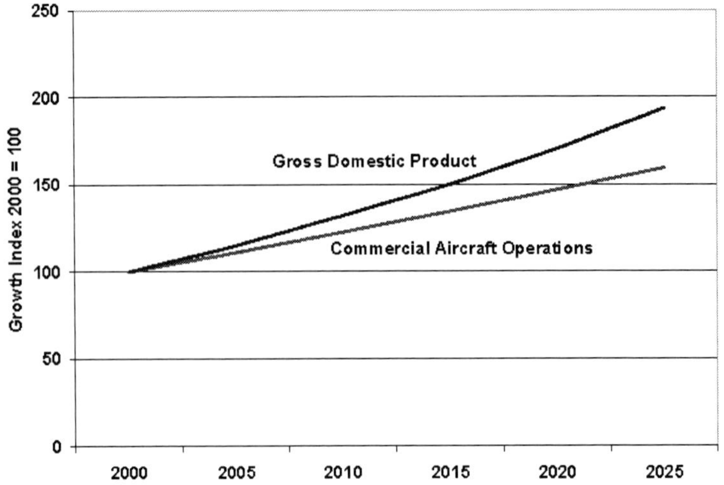

Long range forecasts of aviation activity anticipate continued growth but at a somewhat slower rate than for the U.S. economy as a whole.

As a result, growth of the aircraft fleet and expansion and further development of existing airports are expected. This also means that emissions from aviation activity are expected to grow and concerns about aviation emissions will also grow.

WHAT HAVE BEEN THE TRENDS ON AVIATION EMISSIONS?

Technological advancement has reduced aircraft fuel consumption and emissions significantly over the last 30 years and this is expected to continue in the future.

The industry's historical transition from piston engines to modern high-bypass turbofans resulted in major advancements in energy efficiency and environmental performance. During this same era, the industry developed and deployed new, lightweight, high-strength materials, automated navigational, operational, and engine control systems, and employed vastnew computational capabilities to improveaerodynamic efficiency and integrate highlycomplex operational strategies.

Changes to fleet average fuel economy progressslowly as commercial passenger service aircrafttypically remain in the fleet for 35-40 years. Thenext chart[8] shows the trend and projections inaircraft fuel economy over time.

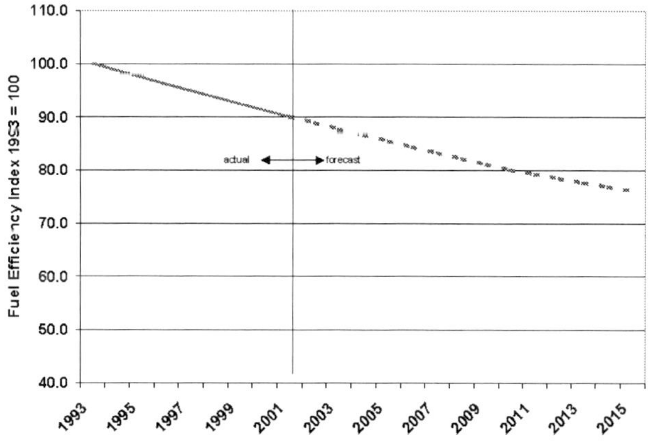

Aircraft fuel efficiency has historically improved by about one percent per year. This trend is expected to continue for the foreseeable future.

As they age, existing aircraft are retired andreplaced with new aircraft. New aircraft also areadded to the fleet for new capacity. By 2020,70% of the fleet will be aircraft added since 2002, which will have advanced technology andcapabilities. The chart below, replicated from anInternational Civil Aviation Organization(ICAO) report[9], illustrates this transition of theinternational commercial aircraft fleet.

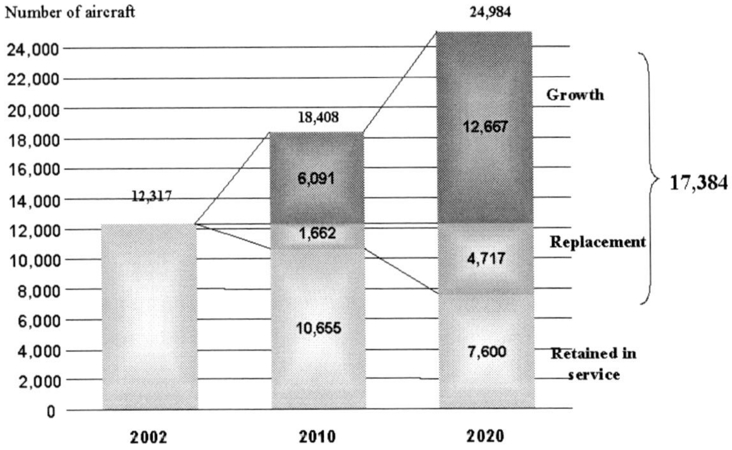

Over an 18-year period, many of today's aircraft will be retired as they reach the end of their life. These aircraft will be replaced and other aircraft will be added to accommodate the growing demand for air travel. The new fleet will be much more energy efficient and have lower emissions.

How Do Aviation Emissions Compare to General Trends in Local Air Pollutants?

Compared to other sources, aviation emissionsare a relatively small contributor to air qualityconcerns both with regard to local air quality andgreenhouse gas emissions. While small,however, aviation emissions cannot be ignored.

In the past three decades, aggregate emissions of the air pollutants EPA regulates (nitrogen dioxide, ozone, sulfur dioxide, particulate matter, carbon monoxide, and lead) have declined by 25 percent nationally, according to their report *National Air Quality 2001 Status and Trends*[10]. As can be seen in the following chart, greater progress has been made with some individual pollutants than with others.

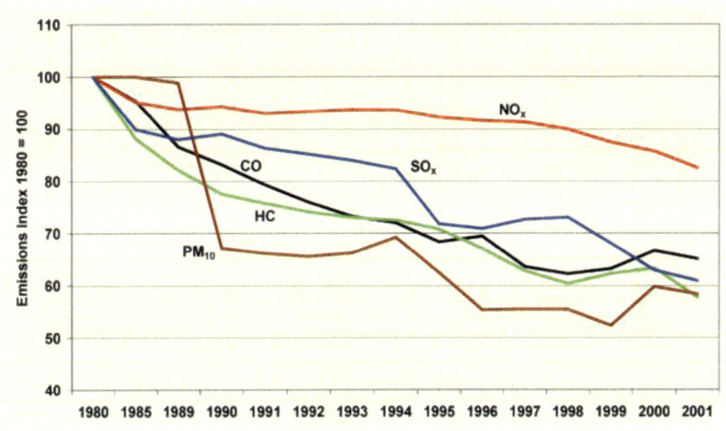

Local air quality pollutants have declined steadily over the past several years. NO_x has been the most challenging pollutant to constrain and progress has lagged that of other pollutants.

Aircraft emissions have also declined over time when you consider the emissions from transporting one passenger one mile. The following chart[11] shows that relative aircraft emissions have fallen consistently over time. Total aircraft emissions have increased, however, since aviation has grown considerably over the same period. As with emissions nationally, a great deal of progress has been made reducing emissions of HC and CO.

NO_X, a key constituent of ozone, has proven to be the most difficult pollutant to control both nationally and for aviation. NO_X comes from a wide variety of sources in all sectors of the economy. Since esssentially all NO_X comes from combustion processes, electric utilities, industry, and transportation are significant emitters and make up the largest share of the total inventory. Currently aviation contributes 0.4 percent of the inventory as can be seen in the illustration[12] on the following page.

Aviation's contribution to the national NO_X emissions inventory has recently declined further as air travel growth has been interrupted during the past two to three years due to the terrorist acts of 9/11, the war on terrorism in Iraq, the emergence of severe acute respiratory syndrome (SARS), and a generally difficult economic environment. These conditions have caused a more than ten- percent decline in air traffic and a similar drop in emissions.

However, these factors are not likely to have a permanent effect on air transportation and growth of travel demand and emissions have recently resumed.

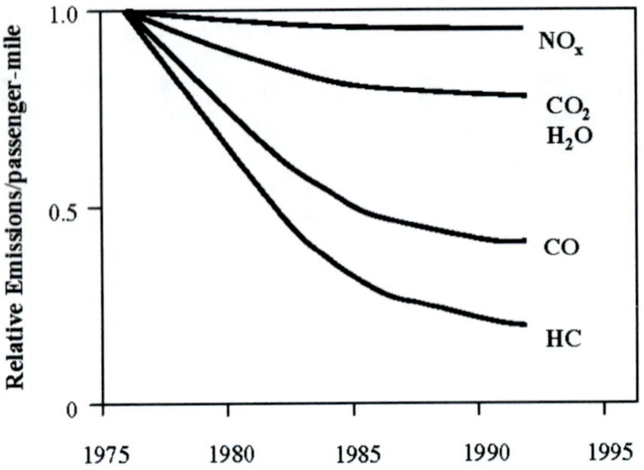

Aircraft emissions of all species have declined over time, however, considerably more progress has been made with HC and CO than with NO_x.

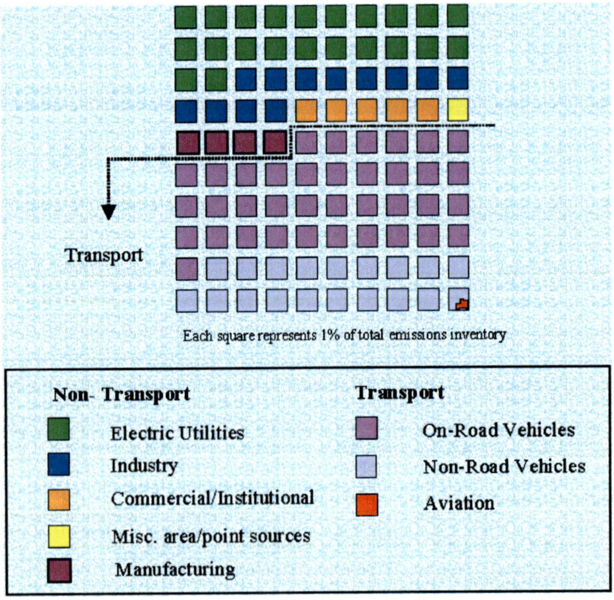

While all transportation makes up more than 55 percent of the total national NO_x inventory, aviation represents only about 0.4 percent.

Total national pollutant inventory numbers do not tell the full story with regard to aviation's contribution in regions with air quality problems. The worst local air quality generally occurs in and around cities, which is also where aviation activity primarily occurs. The Clean Air Act requires EPA to identify air quality areas and to determine whether they comply with (i.e. attain) National Ambient Air Quality Standards[13]. Ozone is by far the principal air quality problem in U.S. cities today. According to EPA data[14], there are currently 474 counties, out of 3,142 counties nationally, that do not meet the new 8- hour ozone standard and are considered nonattainment areas. Comparing this list to the location of primary U.S. airports, 37 of the 50 largest airports are in ozone nonattainment areas.

To calculate aviation's contribution to regional NO_x, we can examine local emission inventories. A multitude of sources comprises air quality area emission inventories.

- *Point sources* – large stationary, industrial facilities that are regulated under Federal, state, or local regulations,
- *On-Road Mobile sources* – cars, trucks, buses, and other vehicles licensed for highway travel,
- *Non-Road Mobile sources* – aircraft, ground support equipment, construction equipment, farm equipment, boats, locomotives, and lawn and garden equipment, and
- *Area sources* – small sources that individually have low emissions but that are significant when combined throughout the area like dry cleaning establishments, bakeries, painting, and vehicle fueling.

The table on the following page summarizes aviation's contribution to NO_x emission inventories in several metropolitan areas. All of these areas have at least one airport that is among the 50 largest airports in the country. To provide a context, Atlanta, Chicago O'Hare, and Los Angeles International are the three busiest U.S. airports. In 2002, Atlanta had nearly 900,000 aircraft operations and enplaned over 37 million passengers; Chicago O'Hare had over 900,000 aircraft operations and enplaned almost 32 million passengers; and Los Angeles International had nearly 800,000 aircraft operations and enplaned almost 27 million passengers.

In the Southern California area, categorized as "severe" nonattainment, EPA's most restrictive designation, aviation's contribution was less than two percent even where the cumulative NO_x from multiple airports was included. While it is apparent from this data that aviation emissions make only a small contribution to regional emissions, even at the largest airports and even in areas with the worst air quality, it is still a contribution that needs to be dealt with effectively.

Airport	**National Rank (enplanements)**[15]	**Ozone Attainment Area Status**[16]	**Airport Contribution to Area NO_x Inventory**	**Aircraft Contribution to Non-Road NO_x Inventory**
Hartsfield Atlanta International (ATL)[17]	1	Marginal	2.8%	14.1%
Chicago Nonattainment Area (ORD, MDW)[18]	2 (ORD), 28 (MDW)	Moderate	0.8-2.0%	10.5%
South Coast California (BUR. LAX, LGB, ONT, SNA)[19]	3 (LAX), 44 (SNA), 51 (ONT), 61 (BUR), 93 (LGB)	Severe	1.5%	5.7%
Dallas/Fort Worth Air Quality Area (DFW, DAL, AFW)[20]	4 (DFW), 53 (DAL)	Moderate	6.1%	19.9%
Houston Bush Intercontinental (IAH)[21]	8	Moderate	0.7%	3.3%
New York (JFK, LGA, EWR)[22]	12 (EWR), 13 (JFK), 21(LGA)	Moderate	4.0%	13.8%
Seattle-Tacoma International (SEA)[23]	15	Attainment	1.9%	6.7%
St. Louis Lambert International (STL)[24]	17	Moderate	1.4%	8.5%
Boston Logan International (BOS)[25]	20	Moderate	0.7%	2.3%

Airports, including aircraft, ground support equipment, and all other vehicles operating around the airport, contribute only a small percentage of NO_x emissions to regional inventories even in cities with the greatest concentration of aviation activity. All of the cities shown include at least one of the 20 largest airports in the country and, except for Seattle-Tacoma, are nonattainment for NO_x under the new 8-hour ground-level ozone designation.
(Inventories are computed from available data for 1996 Atlanta & Dallas/Ft. Worth, 1999 Houston, New York, Seattle & Boston, 2000 Chicago & St. Louis, 2001 South Coast)

HOW DO AVIATION LOCAL EMISSIONS COMPARE TO OTHER TRANSPORTATION SOURCES?

Aviation has grown faster than other modes of transportation and is expected to outpace them in the future. Despite this growth, aviation's contribution to local air quality inventories compared to other transportation sources has remained modest.

For example, as the chart below[26] shows, the *rate of growth* in aviation NO_x emissions has far outpaced on-road (i.e., cars and trucks) NO_x emissions growth.

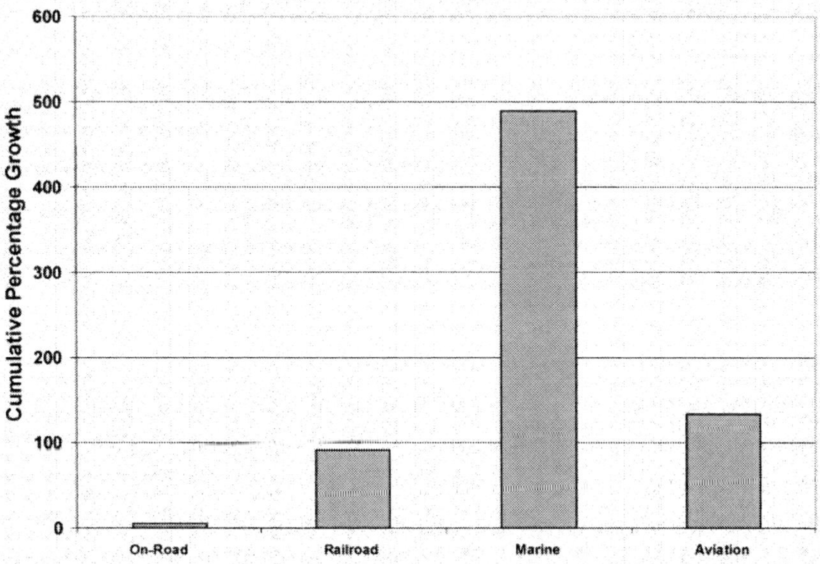

Aviation and marine transportation grew substantially between 1970 and 1998 resulting in large percentage growth in NO_x emissions. Percentage growth in on-road emissions is small due to the large baseline emissions.

However, as the next chart shows[27], the *quantity of emissions* of other transport modes far exceeds aviation's NO_x contribution.

This is not surprising as most of an aircraft's operations take place at altitude where emissions do not affect local air quality. This is in contrast to cars and trucks that primarily operate within a single air quality region and always at ground level[28].

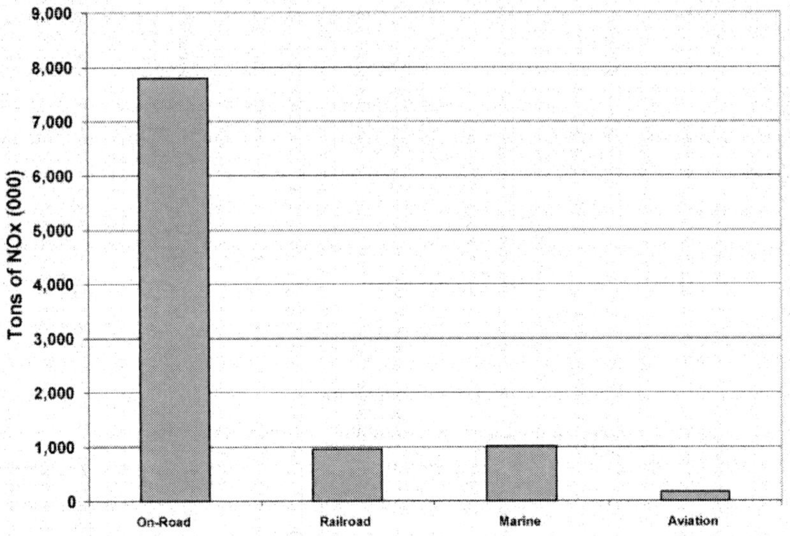

Total NO_x emissions from on-road transportation dwarf emissions from all other transportation modes combined (1998 data).

In the future, NO_x emissions from on-road vehicles should fall in response to the most recent environmental regulations. Aircraft, on the other hand, will be challenged to reduce their total NO_x emissions. However, even if the other sources were able to reduce their emissions by half, a highly unlikely occurrence, aviation NO_x emissions would still be less than 3 percent of the transportation NO_x inventory by 2020. Nonetheless, pressure on aviation sources will likely remain as many states and localities will face the twin challenges of meeting new ozone and particulate matter standards at the same time non-aviation source reductions become more difficult and costly.

Can a Comparison Be Made Between Aviation Emissions and Non-Transportation Sources?

There is an understandable interest in comparing aviation emissions to other sources in local air quality areas. For example, the total mass of emissions coming from an airport may be comparable to those of a power plant or petroleum refinery in the same region.

Airports, however, are quite different from non- transportation sources. Like cities, they are comprised of a variety of different emission sources. Aircraft arrive at the airport, stay for a short period and depart, with a different aircraft taking off or landing every few minutes. Passenger cars, shuttle buses, and taxis calling on the airport do not operate there exclusively, also serving homes and retail, commercial, and governmental establishments. Power boilers and chillers at the airport are independently permitted, as is similar equipment at other locations. For these reasons it is difficult to compare the composite of sources that make up an airport to another emission source like an industrial facility or power plant even when their magnitude of emissions is similar.

What Role Does Aviation Emissions Play with Regard to Greenhouse Gas Issues?

As noted earlier, the majority of aviation emissions occur at higher altitudes, thus generating greenhouse gases and potentially contributing to climate change. Also, under certain conditions, aircraft engine exhaust can produce contrails. Scientists in the U.S. and around the world are researching the potential impact of contrails to see whether they have a significant impact on the greenhouse effect.[29]

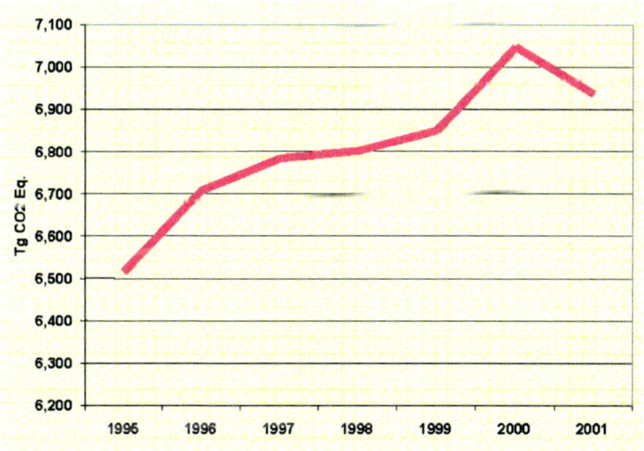

U.S. greenhouse gas emissions grew steadily throughout the 1990s as the economy expanded. The recent downturn is expected to be temporary due to economic recession. With recovery and expansion the upward trend is forecast to continue.

Concern regarding greenhouse gas emissions has been building worldwide. The following graph[30] shows the recent growth of total greenhouse gas emissions in the U.S. The drop in emissions in 2001 reflects the slow economic growth and reduced industrial output that year in addition to the warm winter, which reduced fuel use for heating. Growth in greenhouse gas emissions is expected to resume as the economy recovers and continues to expand in the future. While there are many different greenhouse gases, CO2 and NO_x are generally most relevant from an aviation perspective.

The chart below[31] breaks down national emissions of greenhouse gases. In the U.S., transportation makes up about 27 percent and aviation about 2.7 percent of the national greenhouse gas inventory. Global estimates[32] are similar with emissions of the world's aircraft fleet at about three percent of the total greenhouse emissions from fossil fuel, the majority of which come from commercial aviation. This compares with all transportation sources that contribute approximately 25 percent of total global fossil fuel combustion emissions.

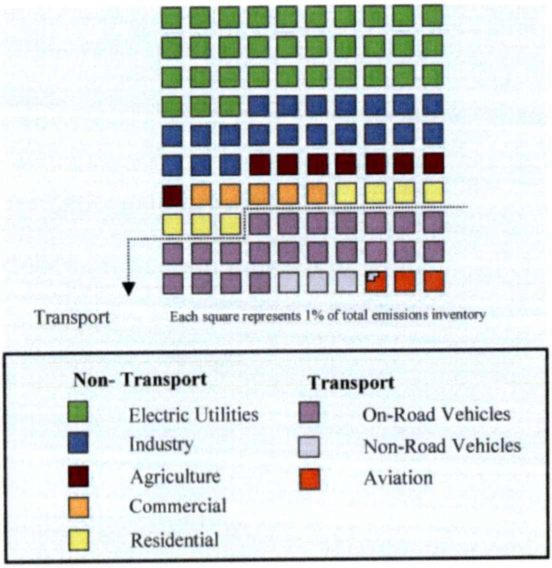

National greenhouse gas emissions in 2001 came from all sectors of the economy with all transportation equal to 27% of the total

Greenhouse gas emissions from U.S. aviation have grown over the past 10 years and are projected to increase in the future[33]. The projection shown below conservatively assumes the relationship between aircraft operations and greenhouse emissions remains constant. As such, emissions track expected growth in aviation. According to the projection, aircraft greenhouse gas emissions in the U.S. will increase 60 percent by 2025[34].

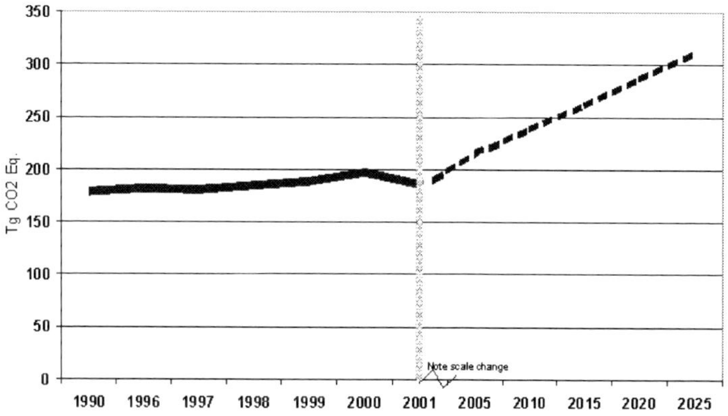

Greenhouse gas emissions from aviation have declined recently due to the fall off in air travel following the terrorist acts of 9/11, the war on terrorism in Iraq, and the worldwide recession. As air travel recovers in the coming years, greenhouse gas emissions are expected to resume their climb.

How Do Aviation's Greenhouse Gas Emissions Compare to Other Transportation Sources?

Energy intensity, that is the amount of energy consumed to transport one passenger one mile, is a useful metric for comparing greenhouse gas emissions among different transportation modes. The different modes use similar fuels and greenhouse gas emissions are directly related to fuel use.

As you can see in the following chart[35], in the U.S., aircraft and automobiles have very similar energy intensities, with automobiles at 3,543 Btu/passenger mile versus airlines at 3,666 Btu/passenger mile. Rail has the lowest energy use, and hence emissions, per passenger mile among all transport modes. Personal trucks and transit buses have the highest energy intensities.

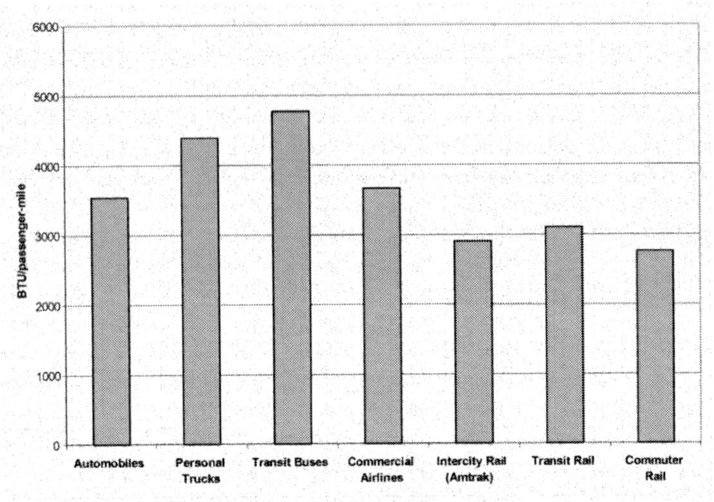

Energy intensity in terms of energy used to carry one passenger one mile is lowest for rail, followed by aircraft and automobiles, with personal trucks and transit buses as the least energy efficient.

Comparing the energy intensity of aircraft and cars shows how energy efficiency and consequently greenhouse gas emissions per passenger mile have changed over the past 35 years. The chart below[36] shows how significant this has been, especially for aircraft.

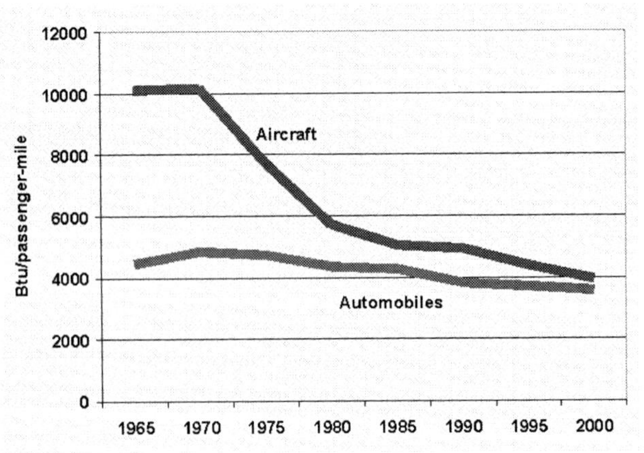

The energy intensity of aircraft and automobiles has improved substantially over the past several decades. Automobile energy intensity has fallen by almost one fifth while aircraft energy intensity has fallen by three fifths during the same period.

The pie chart[37] at the top of the following page shows the total energy consumption for each transportation mode. Since the fuels are similar, this is an indication of their total greenhouse gas emissions. Aviation is substantially less than automobiles and personal trucks though significantly more than rail and buses. However, given the greater potential to apply alternative fuel technologies to land-based transport in the next two decades, aviation greenhouse gas emissions are likely to represent a greater share of transport sources over time.

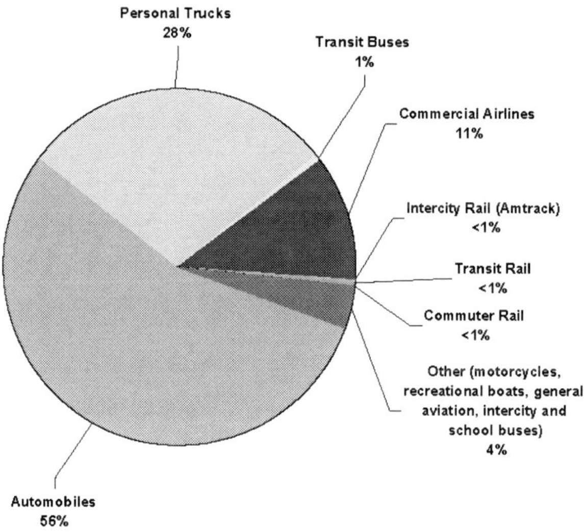

Energy consumption by the aviation industry is only a small portion of transportation energy use. Automobiles and personal trucks still account for the vast majority.

How are Aviation Emissions Regulated?

There is some misperception that aviation in general and airlines in particular are the "only unregulated industry in the country," or "are getting a free ride on air quality," and "cars have reduced their emissions by over 98% while aircraft have done nothing." In fact, there are many, varied regulations that constrain aviation emissions. For example, both cars and aircraft have improved their energy intensity over time using new technologies, advanced materials, and improved designs for energy conservation to reduce fuel consumption.

Practically all aviation emission sources are independently regulated through equipment-specific regulations, standards and recommended practices, and operational guidelines, which are established by a variety of organizations. For example, on-road vehicles, which take passengers to and from the airport, meet stringent Federal tailpipe standards set by EPA. Stationary sources on the airport, like power boilers and refrigeration chillers, must meet independent state regulations. And FAA certification is required for essentially all aviation equipment and processes. For example there are more than 60 standards[38] that apply to aircraft engine design, materials of construction, durability, instrumentation and control, and safety, among others. These are in addition to the Fuel Venting and Exhaust Emission Requirements for Turbine Engine Powered Airplanes (FAR Part 34), which guide compliance with EPA's aircraft exhaust emission standards. This comprehensive and complex regulatory framework has enabled our safe and efficient national air transport network.

The International Civil Aviation Organization (ICAO) is a United Nations intergovernmental body responsible for worldwide planning, implementation, and coordination of civil aviation. ICAO sets emission standards for jet engines. These are the basis of FAA's aircraft engine performance certification standards, established through EPA regulations.

ICAO has long been the forum for evaluating the environmental performance of aircraft engines. ICAO has taken a "technology progressing" approach, raising standards within the capabilities of proven technologies and certified products (engines and aircraft) rather than a "technology forcing" approach, which sets standards based on technology that is not certified or may not even exist. The reason for ICAO's approach is quite simple - the very high premium placed on the safety of aircraft operation restricts the use of unproven new technologies.

Current NO_x standards were established in 1996. New standards go into effect for engines entering service beginning in 2004, which reflect a 16 percent NO_x reduction over the 1996 standards and a 33 percent reduction over the original standards agreed to in 1981. Earlier this year, ICAO's Committee on Aviation Environmental Protection recommended new certification standards that represent a further 12 percent NO_x reduction, with an effective date of 2008.

Airport air emissions from all sources also are constrained by the General Conformity regulations of the Clean Air Act Amendments of 1990. General Conformity requires Federal agencies to assure that actions that would increase emissions in nonattainment areas "conform" to the appropriate State

Implementation Plan. These plans define the steps states are committed to taking to ensure their cities enjoy healthy air. Each year the environmental impacts of several hundred projects at airports throughout the country are analyzed in detail, including general conformity evaluations and analyses, using the best data and most advanced analytical models available. Emissions from the vast majority of these projects are well below the thresholds that trigger a "conformity determination." The two or three projects a year that do require further analysis essentially are able to meet the needs of state air quality plans through minor project modification.

EPA recently proposed new exhaust emission standards for non-road diesel engines. These standards, to be phased in between 2008 and 2014, will require engine manufacturers to produce new engines with advanced emission control technologies. New ground support equipment with diesel engines, which are used only on airport property, will be required to meet these standards. This new equipment will achieve emission performance comparable to today's automobiles.

While there are no national or international regulations for greenhouse gas emissions that apply to aircraft or other airport sources, the aviation industry has made significant strides here as well. Aircraft have a long history of continuously improved fuel economy, which reduces all greenhouse gas emissions. For example, according to Boeing, the B-777 is 300 percent more efficient than its early jets.[39] Fuel economy and energy conservation are also priorities at many airports. Dallas/Fort Worth International Airport for example looks at business practices at all of their facilities to minimize energy consumption. They realize that this benefits local air quality through reduced emissions as well as regional air quality as a result of reduced power purchases from electric utilities and an overall reduction in greenhouse gas emissions.

Looking to the future, FAA is working through ICAO to evaluate policy options to limit or reduce greenhouse gas emissions from aviation. Various market-based options, such as voluntary agreements, open emissions trading, and emission related levies are being analyzed. Preliminary results from analyses of market- based options show that emission related levies are not cost-beneficial, but voluntary arrangements and emissions trading may be cost effective in limiting or reducing greenhouse gas emissions. Additional analyses are underway at ICAO to evaluate further emissions trading and voluntary agreements as approaches to limit aviation emissions growth while allowing continued expansion of air travel[40].

Under this multidimensional regulatory and voluntary structure, aviation has made significant environmental progress. Given the complexity of the industry and the need for different strategies and technological approaches for different types of vehicles and equipment, a coordinated effort between the aviation industry and the many regulatory agencies that share environmental responsibility will continue.

What is Being Done Today to Reduce Aviation Emissions?

There are a number of initiatives underway that will achieve significant emissions reductions – both at airports and within the national aviation system – in the next few years. First, there are voluntary programs underway at airports to reduce emissions from ground support equipment and other airport vehicles. For example, FAA developed a pilot program, with EPA and DOE, to demonstrate air quality improvements with alternative fuel ground support equipment. The program is called the Inherently Low-Emissions Airport Vehicle (ILEAV) Pilot Program.

To reduce emissions from these vehicle fleets, airlines have engaged in voluntary emission reduction programs. For example, California and Texas have agreements with the major airlines to reduce emissions from their ground support equipment. These new agreements will reduce emissions by converting gasoline and diesel equipment to electricity and alternative fuels. A national stakeholders group made up of representatives of FAA, EPA, major airlines, state and local environmental regulators, airports, and environmental interest groups is currently working to establish a national agreement to reduce ground support equipment emissions at other airports in air quality nonattainment areas. This has proved challenging, and it is still unclear whether it will be successful.

In addition, many airports have independently taken action to reduce emissions from buses, trucks, taxicabs, and other on-road vehicles that operate in and around the airport. Hybrid-electric vehicles are being used for staff transportation and customer service vehicles. Airport police departments are using compressed natural gas automobiles and maintenance departments are using alternative fuel trucks. Airport shuttle buses in particular have been converted to compressed natural gas at a number of airports. Also, new clean diesel trucks are being used in heavy maintenance and construction.

Based on their experience with the ILEAV Program, FAA and EPA have expanded the initiative to reduce ground emissions at commercial service airports in all air quality nonattainment areas. The new Voluntary Airport Low Emission (VALE) program expands eligibility for airport low emission projects under the Airport Improvement Program (AIP) and the Passenger Facility Charges (PFC) program. Through the use of funding and emission credit incentives, the voluntary program includes the conversion of airport vehicles and ground support equipment to low emission technologies, modification of airport infrastructure for alternative fuels, provision of terminal gate electricity and air for parked aircraft, a pilot program to explore retrofit technology for airport ground support equipment, and other related emissions improvements.

Second, it is also worth noting that many strategies for reducing the environmental impact of aviation are inherent to the intended design and operation of the air transport infrastructure. With airports for example, access roadways are often limited access, high-speed and free flowing and parking facilities are readily available. These features minimize motor vehicle emissions and keep them contained in areas away from the public.

Third, looking at aviation more broadly, many recent changes have improved the system efficiency and reduced environmental impact. In the past few years, better meteorological information, available in the cockpit in real time, has allowed for optimized flight planning with shorter routing. Yield management tools have allowed airlines to increase load factors, which moves more people on every flight. The hub and spoke system, combined with the growth of low cost point-to-point carriers and a significant increase in the number and reach of regional airlines, has improved the efficiency of the entire aviation network.

Fourth, operating procedures can have both direct and indirect effect on aircraft emissions. Airlines generally employ standard procedures for operating their aircraft to meet company goals for safety, adherence to flight schedules, fuel conservation, complying with labor agreements, and other factors. Standard procedures vary by aircraft type, airport-specific constraints, and weather. The use of alternative procedures or best practices offers some prospect for reducing emissions.

Some procedures affect the engine-operating regime, which can directly influence the rate of pollutant emissions. NO_x emissions are higher during high power operations like takeoff when combustor temperatures are high. On the other hand, HC and CO emissions are higher during low power operations like taxiing when combustor temperatures are low and the engine is less efficient. As a result, reducing engine power for a given operation like takeoff or climb out generally increases the rate of HC and CO emissions, reduces the rate of

NO_x emissions, and has little or no effect on CO2 emissions. Other operating procedures have a more general effect on engine use and can reduce all pollutants simultaneously.

As another example of alternative operating practices at an airport, United Airlines launched a new initiative last year to reduce the average use of its auxiliary power units by using ground power whenever possible. Based on early tests of the program they expect to save approximately 12 million gallons of fuel during the year, which will result in reduced emissions of all pollutants at the airport as well[41]. Many of the strategies discussed in this section are published in ICAO Circular 303 - *Operational Opportunities to Minimize Fuel Use & Reduce Emissions*[42].

New technologies to improve air traffic management will help reduce emissions in and around airports. Commonly referred to as CNS/ATM (communication, navigation, surveillance/air traffic management), many of these technologies will improve air traffic management efficiency in the terminal area air space, reduce congestion, and consequently reduce aircraft fuel use. They will ensure more accurate approach routes to precisely keep aircraft on track. They will increase the efficiency and capability of runways, reducing arrival spacing and making ground operations more efficient. These systems are expected to be in operation throughout the U.S. over the next 10-15 years[43].

A near term example is RVSM – Reduced Vertical Separation Minimums. Reducing vertical separation between aircraft from 2000 feet to 1000 feet separation at cruise altitude (i.e., above 29,000 feet) adds flight levels and increases airspace capacity by as much as 85%. These routes are among the most fuel efficient for long flights such as oceanic or cross-country traffic and increasing their availability allows for greater flexibility in flight scheduling and routing. RVSM has been in use for transatlantic flights since 1997 and will become standard in U.S. domestic airspace starting in January 2005. Fuel savings of more than 500 million gallons each year are expected in U.S. airspace alone with full implementation of RVSM[44].

What Steps Are Being Taken to Reduce Aviation Emissions in the Longer-Term?

The pace of technological change across the industry is increasing. New engine designs are improving fuel efficiency further, while simultaneously reducing NO_x emissions. New aircraft designs improve aerodynamics and reduce weight thereby improving fuel efficiency, reducing all pollutants at the same time. New air traffic control technologies, like new aircraft designs, reduce emissions by reducing fuel consumption (the effect on individual pollutants depends on the phase of flight most effected). And new management strategies like load management planning and code sharing are being used to optimize the entire system's operation.

With regard to engines, there are complex emission interrelationships that make it difficult to modify their design as a mitigation strategy since it forces a tradeoff among individual pollutants as well as between emissions and noise. For example, high-bypass turbofan engines were introduced to reduce noise and improve fuel efficiency. They require higher engine pressure ratios, which increase engine temperatures, and hence generate more NO_x. It has only been in the past 25 years that the resulting NO_x increase became a concern. FAA and other stakeholders have recently initiated an effort to better understand and quantify these interrelationships in an "environmental design space." Eventually this will lead to guidelines for setting long term goals and standards that optimize overall environmental performance and avoid unintended consequences.

Aircraft design improvements mostly fall into one of three areas: aerodynamics, weight reduction, and control systems. Continued improvement in all areas is expected in the future[45]. Some of the technologies in development for aerodynamic improvements include the design of winglets for wing tips, which reduce turbulence and vortex generation by the wings, laminar flow controls or systems for wing surfaces to reduce drag, and improved manufacturing techniques that will produce smoother surfaces. New and improved metal alloys and composite materials are being developed to reduce aircraft weight while simultaneously improving structural performance. Significant improvement of control systems has come about by replacing mechanical and hydraulic systems with electrical systems, which often reduce system weight while providing more precise control. Improvements of these systems and development of new systems for enhanced flight stability will contribute to improved overall fuel efficiency.

Aircraft technology development and capital turnover follow relatively long cycles, which limits the pace of fundamental changes in design. It takes approximately 10 to 15 years for fleet average fuel efficiency to equal the efficiency of the newest aircraft[46]. However, the ongoing evolutionary change in technology has realized substantial benefits over time. According to the Intergovernmental Panel on Climate Change, in their report *Aviation and the*

Global Atmosphere[47], aircraft fuel efficiency has improved by 75 percent in the past 40 years through improvements in airframe design, engine technology, and rising load factors.

While progress in the near- and mid-terms is expected, the most significant opportunities for emission reduction lie in the future when we can derive benefits from aggressive research goals. FAA, EPA, ICAO, and many other groups have been working to elucidate and characterize the environmental issues for some time while NASA is directing a research program aimed at significantly cutting emissions from aircraft engines[48].

A comprehensive research program starts with a clear and complete understanding of the effects of aviation on air quality. FAA has developed analytical tools to quantify emissions more accurately, which are used to understand aviation's contribution to local air quality concerns and global emissions. These tools are developed on a foundation of research that FAA has conducted both independently an in conjunction with airports and other organizations. The Emissions and Dispersion Modeling System (EDMS) has been developed to quantify emissions from aircraft and other airport emission sources. It is used routinely to assess the impact of airport expansion projects and other operational changes. The System for assessing Aviation's Global Emissions (SAGE) is being developed to assess the impact of aircraft engine emissions during the whole flight regime, especially climb out and cruise emissions. The model will be able to develop aviation emission inventories, both for baseline conditions and forecasted technology, and assess operational and market-based measures and improvements. New tools are also being developed to understand and assess the environmental design space, to evaluate interrelationships among all emissions and between emissions and noise due to changes in technology and operational procedures.

NASA is the U.S. federal agency responsible for pre-commercial aerospace research, development, and demonstration. One of the key themes of their research program is to "protect local environmental quality and the global climate by reducing aircraft noise and emissions." Their strategy is to research opportunities to reduce airframe weight and drag, optimize engine systems, and optimize operations at and around airports. Through this, NASA hopes their research program results in significant or total elimination of aircraft greenhouse gas emissions, minimized impact of emissions on local air quality, and elimination of unnecessary aviation emissions due to operational procedures. The goals of this program are to reduce NO_x emissions of future aircraft by 70 percent by 2007, and by 80 percent beyond 2007 using 1996 ICAO standards as a baseline. They also intend to reduce CO2 emissions of future aircraft by 25 percent and by 50 percent for these same milestones using 2000 state-of-the-art aircraft technology as a baseline[49].

The primary engine research project to achieve these objectives is the Ultra-Efficient Engine Technology (UEET) project, in NASA's Vehicle Systems Program. According to NASA, the UEET project will develop and transfer to U.S. industry critical turbine engine technologies that will contribute to enabling a safe, secure, and environmentally friendly air transportation system. This project is currently underway.

FAA, NASA, and Transport Canada have made a major commitment to researching aviation emissions as well as noise through the Center of Excellence for Aircraft Noise and Aviation Emissions Mitigation.[50] The Center was established in September 2003 to foster breakthrough technical, operational, and workforce capabilities enabling quieter and cleaner aircraft.

Achieving research goals will allow the aviation industry to significantly reduce its environmental impact and begin to reduce its total emissions of NO_x and CO2. This takes time, however. As noted earlier it takes 10 to 15 years for fleet average performance to achieve current new technology performance. To go from NASA research to fleet average performance takes 20 to 40 years[51]. Also, concerns have been raised about the likelihood of these goals being met due to budget restrictions. The National Research Council recently published a report, *For Greener Skies: Reducing Environmental Impacts of Aviation*[52], that concluded, while "the goals of the federal research program are admirable and focused on the right issues, the schedule for achieving the goals is unrealistic in view of shrinking research budgets." The report went on to call for further federal investment in engine research and technology development.

Aviation Emissions Are Being Responsibly Controlled

Aviation has progressively improved its environmental performance. Fuel economy, which is one strong indicator of environmental performance, has consistently improved. Aircraft engines have gotten more efficient and been designed with environmental performance in mind. Regulatory frameworks have developed to constrain emissions growth from many aviation sources. And improvements to the efficient operation of the complex aviation network have had a positive effect on the environment.

Looking to the future, FAA has a roadmap for continuing to mitigate the environmental impacts of aviation. This includes continuing to improve its understanding of the role of aviation emissions on the environment. FAA is working with industry and other stakeholders to advance the performance of the national and international aviation system as well as to improve individual system components. And FAA is working in the international arena to evaluate alternative strategies for market-based opportunities for reducing emissions.

FAA, together with EPA and NASA, is committed to ensuring aviation emissions do not pose health concerns for our citizens or restrain aviation's mobility and economic benefits enjoyed by society. It will take consistent, coordinated effort and continuing success in technology research and development to achieve these goals.

End Notes

[1] Population Timeline, http://www.pbs.org/kqed/population

[2] http://www.atag.org/files/FAST%20FACTS-120341A.pdf

[3] *Ibid*

[4] http://www.iata.org

[5] U.S. Department of Transportation, Bureau of Transportation Statistics, *Transportation Indicators* http://www.bts.gov/publications/transportation, December 2002.

[6] U.S. Department of Transportation, Federal Aviation Administration, *FAA Long-Range Aerospace Forecasts Fiscal Years 2015, 2020 and 2025*, Office of Aviation Policy and Plans, FAA-APO-00-5, http://apo.faa.gov/lng00/lng00.pdf, June 2000.

[7] GDP - U.S. Department of Commerce, Bureau of Economic Analysis, *National Accounts* Data http://www.bea.gov/bea/dn1.htm; RTM - U.S. Department of Transportation, Bureau of Transportation Statistics, *Historical Air Traffic Statistics*, I2001html; VMT – U. S. Department of Transportation, Federal Highway Administration, *Traffic Volume Trends*, December 2002 http://www.fhwa.dot.gov /ohim/ tvtw/02dectvt/tvtdec02.pdf.

[8] Ibid. Table 9-15.

[9] Wickrama, Upali, International Civil Aviation Organization, Committee on Environmental Protection, Forecasting and Economic Analysis Support Group, *Report of the FESG/CAEP/6 Traffic and Fleet Forecast*, copy of Figure 8, 2003 op. cit.

[10] U.S. Environmental Protection Agency, *National Air Quality 2001 Status and Trends*, http://www.epa.gov/airtrends/aqtrnd01/, September 2002. [http://www.epa.gov/air/airtrends /aqtrnd03/fr_table.html]

[11] Waitz, I. A., Massachusetts Institute of Technology, private communication based on Boeing data, November 2003.

[12] U.S. Environmental Protection Agency, *Average Annual Emissions, All Criteria Pollutants; Years Including 1980, 1985, 1989-2001*, http://www.epa.gov/ttnchie1/trends/index.html, February 2003.

[13] Clean Air Act Amendments of 1990, Title I – Provisions for Attainment and Maintenance of National Ambient Air Quality Standards, Section 101(d)(1), November 15, 1990.

[14] U.S. Environmental Protection Agency, *8-Hour Ground-level Ozone Designations*, http://www .epa.gov/ozonedesignations/statedesig.htm, May 6, 2004.

[15] U.S. Department of Transportation, Federal Aviation Administration, *Enplanement Activity at Primary Airports*, http://www.faa.gov/arp/planning , November 6, 2003.

[16] U.S. Environmental Protection Agency, *Classifications of Ozone Nonattainment Areas*, op.cit.

[17] U.S. Department of Transportation, Federal Aviation Administration, Federal Highway Administration (cooperating agency), *Final Environmental Impact Statement for 9,000-Foot Fifth Runway and Associated Projects: Hartsfield Atlanta International Airport*, August 2001.

[18] Illinois Environmental Protection Agency, *Illinois 1999 Periodic Emissions Inventory And Milestone Demonstration,* December, 2001. The higher value for in the area inventory data in the table is for a typical summer day, which is the ozone season and probably represents a worst case since it is the most active period for aviation activity. The non-road data also is based on typical summer day. The lower value, which is more representative for an annual value is from U.S. Department of Transportation, Federal Aviation Administration, *Final Environmental Assessment for the World Gateway Program and Other Capital Improvements Chicago O'Hare International Airport, Chicago, Illinois*, June 21, 2002.

[19] South Coast Air Quality Management District, *Emissions by Category, 2001 Estimated Annual Average Emissions, South Coast Air Basin.*http://www.arb.ca.gov/app/emsinv/emssumcat_ *query.php?F_DIV=0&F_YR=2001&F_AREA=AB&F_AB* =SC , 2001.

[20] Texas Natural Resource Conservation Commission, *Dallas/Fort Worth Ozone Nonattainment Area Emission Data*, http://www.tnrcc.state.tx.us/air 1996 inventory data. Data includes all airports in the nonattainment area including, DFW International Airport, Dallas Love Field, and Alliance Airport.

[21] U.S. Department of Transportation, Federal Aviation Administration, *Final Environmental Impact Statement Runway 8L-26R and Associated Near-Term Master Plan Projects; George Bush Intercontinental Airport/Houston*, July 2000.

[22] Compilation of data from the SIP inventories for New York and New Jersey provided by Mr. Raymond Forde, Region 2, U. S. Environmental Protection Agency, June 16, 2004. Additional data provided by Mr. Kevin McGarry, New York State Department of Conservation and Ms. Tonalee Key, New Jersey Department of Environmental Protection.

[23] Agyei, Kwame, Puget Sound Clean Air Agency, airport emissions calculated using EDMS 4.0; area non- road and total emissions from 1999 Air Emission Inventory Summary spreadsheet, February 11, 2003.

[24] Nonattainment area non-road and total NO_x emissions, 68 FR 25431, May 12, 2003; Airport emissions escalated from 1995 estimate by URS Greiner, Inc. (1997) based on 2000 data provided by Tony Petruska, U.S. EPA.

[25] Massachusetts Department of Environmental Protection, *Massachusetts Periodic Emissions Inventories 1999*, April 2003, for nonattainment area off-road emissions and total emissions, which are based on summer day emissions. U.S. Department of Transportation, Federal Aviation Administration, *Final Environmental Impact Statement, Logan Airside Improvements Planning Projects: Boston Logan International Airport*, June 2002 for Logan Airport emissions, which are typical for an annual value.

[26] U.S. Environmental Protection Agency, *National Air Pollutant Emission Trends, 1990-1998*, http://www.epa.gov/ttn/chief/trends/trends98/index.html, March 2000.

[27] Ibid.

[28] For NO_x, aircraft represent anywhere from 60 to 80 percent of total airport emissions with the balance coming from the other sources like ground support equipment and ground access vehicles. This is based on a review of recent Environmental Impact Statements for eleven airports (ATL, BOS, CLE, DFW, IAD, IAH, LAX, ORD, PTI, SFO, and STL).

[29] For more information about the production of contrails by aircraft, see U.S. Environmental Protection Agency, *Aircraft Contrails Factsheet*, http://www.epa.gov/otaq/regs/nonroad/ aviation September 2000.

[30] U.S. Environmental Protection Agency, *Inventory of U.S. Greenhouse Gas Emissions and Sinks: 1990- 2001, http://yosemite.epa.gov/oar/globalwarming.nsf/UniqueKeyLookup/*

LHOD5MJQ6G/$File/2003- final-inventory.pdf, April 15, 2003. Estimates are presented in units of terragrams of carbon dioxide equivalents (Tg CO2 Eq.), which weight each gas (e.g., CO2 and NOx) by its Global Warming Potential, or GWP, value.

[31] U.S. Environmental Protection Agency, *Inventory of U.S. Greenhouse Gas Emissions and Sinks: 1990- 2001*, 2003 op.cit.

[32] Intergovernmental Panel on Climate Change, *Aviation and the Global Atmosphere*, 1999.

[33] Actual Emissions 1990-2001 – U.S. Environmental Protection Agency, *Inventory of U.S. Greenhouse Gas Emissions and Sinks: 1990-2001*, 2003 op.cit.

[34] Forecast Emissions 2001-2025 calculated based on FAA long-range activity forecasts assuming a constant rate of emissions from aircraft. The forecast is deemed conservative since it does not account for improvements in aircraft energy efficiency over the next 20 years, which are deemed likely. Estimates are presented in units of terragrams of carbon dioxide equivalents (Tg CO2 Eq.), which weight each gas (e.g., CO2 and NOx) by its Global Warming Potential, or GWP, value.

[35] U.S. Department of Energy, Oak Ridge National Laboratory, *Transportation Energy Data Book: Edition 22*, http://www-cta.ornl.gov/data/tedb22/Full_Doc_TEDB22.pdf, September 2002.

[36] US Department of Transportation, Bureau of Transportation Statistics, *National Transportation Statistics 2002* (BTS 02-08), Table 4-20: Energy Intensity of Passenger Modes (Btu per passenger-mile), page 281, http://www.bts.gov/publications/national_ transportation_statistics/2002/pdf/entire.pdf.

[37] Ibid.

[38] See FAR Part 33 – Airworthiness Standards: Aircraft Engines http://www.airweb.faa.gov /Regulatory_and_Guidance_ Library/rgFAR.nsf/CurrentFAR*Part*?OpenView&St art=1 &Count=200&Expand=1 0.

[39] Colpin, J. and Altman, R., *Dependable Power Reinvented*, AIAA 2003-2882, AIAA-ICAS International Air and Space Symposium and Exhibit: The Next 100 Years, July 14-17, 2003, Dayton, OH.

[40] The use of market-based options for limiting or reducing greenhouse gas emissions from aviation may also prove potentially useful applied to local air quality emissions.

[41] Air Transport World, *United says cutting APU runtime to save $12 million*, ATW Online.com, February 10, 2003.

[42] International Civil Aviation Organization, Circular 303 - *Operational Opportunities to Minimize Fuel Use & Reduce Emissions*, February 2004.

[43] U.S. Department of Transportation, Federal Aviation Administration, *National Airspace System Operational Evolution Plan*, December 2002

[44] U.S. Department of Transportation, Federal Aviation Administration, *Final Regulatory Impact Analysis, Final Regulatory Flexibility Determination, Unfunded Mandates and Trade Impact Assessment, Reduced Vertical Separation Minimum Operations in United States Domestic Airspace*, March 10, 2003

[45] Drew, P., et al., Technology Drivers for 21st Century Transportation Systems," AIAA 2003-2909, AIAAICAS International Air and Space Symposium and Exhibit: The Next 100 Years," 14-17 July 2003, Dayton, OH

[46] Waitz, I. A., Massachusetts Institute of Technology, *Aircraft, Gas Turbine Engines and Emissions Primer*, August 3, 2001.

[47] Intergovernmental Panel on Climate Change, 1999 op.cit.

[48] ICAO has established a Long-Term Technology Goals (LTTG) task group within Working Group 3 to monitor and track future aircraft technologies that may demonstrate better environmental performance. The LTTG will evaluate the prospects for setting emissions goals as targets for future technology performance.

[49] National Aeronautics and Space Administration, *NASA Aerospace Technology Enterprise Strategy – 2003*, http://www.aerospace

[50] More information on the FAA-NASA Center of Excellence for Aircraft Noise and Aviation Emissions Mitigation can be found at http://web

[51] Waitz, I. A., Massachusetts Institute of Technology, 2001 op. cit., estimates "22 to 37 years total time from basic technology (e.g. NASA research) to significant fleet impact."

[52] National Research Council, Division on Engineering and Physical Sciences, Aeronautics and Space Engineering Board, Committee on Aeronautics Research and Technology for Environmental Compatibility, *For Greener Skies: Reducing Environmental Impacts of Aviation*, available at http://bob.nap.edu/html/greener_skies/notice.html, 2002.

In: Aviation and Climate Change
Editors: George T. Blumenthal
ISBN: 978-1-60876-757-1

Chapter 3

HEARING ON AVIATION AND THE ENVIRONMENT: EMISSIONS

Richard L. Altman

Thank you Mr. Chairman and members of the subcommittee for providing the Commercial Aviation Alternative Fuels Initiative with the opportunity to testify on the compelling issues of Aviation and the Environment: Emissions.

Mr. Chairman thank you as well for acknowledging Aviation's efforts in alternative fuels in particular. In your April 8 Op-ed for "The Hill" you stated that "the (aviation industry) is leading the way in research on alternative fuels. Besides the positive impact on the bottom line, there are obvious positive environmental impacts from these efforts, with lessons for the rest of the country". It is particularly gratifying to us in the Aviation sector that, our positive efforts are being recognized including, but not limited to, the efforts of the CAAFI collaboration. CAAFI is solely an instrument that Commercial Aviation is using to bring together the full scope of subjects needed to speed the introduction of alternative fuels. In testifying today as a representative of the CAAFI my goal is to provide new information to the committee on our progress since we last spoke to you one year ago. With progress there also comes added insight to the challenges that are being faced. Those will be addressed as well.

For those not familiar with CAAFI, the Commercial Aviation Alternative Fuels Initiative is a data gathering and communications collaboration that seeks to increase both the quantity and the quality of dialogue among its

Airline, Airport, Manufacturer and FAA sponsors and between sponsors and the numerous stakeholders that are engaged with us. CAAFI engages multiple government, industry and university contributors. Over 20 Energy suppliers are now stakeholders in CAAFI. We are global in reach with multiple stakeholders on four contents. The fundamental belief of the sponsors in forming the CAAFI collaboration is that the aviation industry is data driven and relatively small in size allowing it to benefit from such an approach. CAAFI's sponsors and stakeholders recognize that data they develop and collect needs to be placed in the hands of key analysts and decision makers. In such matters as safety, security, and the environment such analysis will be a catalyst for informed and expedited solutions that serves all components of the supply chain well. Uniform understanding of solutions and clarity of message, it is believed, will spur suppliers to invest in solutions suitable for Commercial Aviation. The goal is to make our relatively small sector of transportation a "customer of choice" for introducing alternative fuels.

With this goal in mind this testimony focuses on two areas. First, progress we have been made in each of CAAFI's four functional areas (Certification, R&D, Environmental and Business/Economics) since we met last year with the committee will be laid out. Second, the main challenges that we face to strengthen and continue our leadership roles in alternative fuels will be highlighted.

As indicated last year the input provided is very much a snapshot of unfolding events. New results arrive and new ideas are conceived almost daily in this rapidly developing area — just as fuel price went up $10 per barrel crude since you authored the Op-ed piece on April 8.

The first four months of 2008 have seen an unprecedented level of activity in the Certification and R&D developments globally. Obviously there are many headline developments you cited in your comments on highly visible Flight programs. Those occurred both at Boeing and at Airbus. Many more are planned.

Most important, in my view, is the spadework being done in three areas. These efforts do not typically draw headlines but will result in turning these headlines into actual fuel production that will affect both producers and users. Accordingly this testimony addresses the efforts of CAAFI teams seeking to

- Execute R&D and certify new fuels
- Define their environmental benefits and costs of these fuels as well as to establish tools to ease the assessment of the benefits of

introducing these fuels to new location using a full life cycle (well to wake analysis).

- Establish efforts to facilitate dialogue between new suppliers and buyers to ensure that the needed discussions take place. This dialogue is critical to ensure potential investors that projects that supply alternative fuels via the unique airport fuel distribution infrastructure will find ready buyers.

Last year CAAFI identified broad time frames when key R&D and certification events would occur. That promise is becoming reality with several truly key developments having occurred in the last few months.

On April 11 the Air Force Certification Office approved a new JP8 Mil Spec (8133) embracing 50/50 blends of Synthetic Paraffinic Kerosene (SPK) from all forms of feed stocks including coal (CTL), gas (GTL), and biomass (BTL). This qualification is the first critical break from the long-term certification approach which qualified fuels on an individual producer/ individual fuel basis. This "generic approval" does much to strengthen the opportunities for new Suppliers.

In February the CAAFI Certification and Qualification team was a driving force behind the submittal of a ballot to the ASTM proposing a modification to the fuel specification ASTM D-1655 to include 50/50 SPK based blends. In a meeting just last week in Alexandria, VA the team mapped out its plans to achieve a positive outcome that is expected to parallel the Air Force process and targets completion by yearend.

Taken together these developments form critical signposts to investors in synthetic jet fuel plants. Aviation fuels can comprise up to 70% of the output from plants like those in the planning stages for construction Illinois and Ohio by Rentech, American Clean Coal Fuels and BAARD Energy. If initial approvals for these projects are granted this year, construction can start early next year and initial production is possible in the 2012 time period.

Promising developments do not stop there. Also in April the UK certification panel provided its approval (via DefStan 9191) for 100% Sasol CTL from Sasol's South African facility. That approval marks a notable success for the Engine manufacturers GE, Honeywell, Pratt & Whitney and Rolls Royce as well as Sasol. Together they were able to run all tests required by a new set of engine company protocols and become certified in about 3 years time. Certification was achieved less than 15 months after the completion of the last test at P&W in January 2007. This is a great sign that we will be able to achieve CAAFI's goal of approving 100% FT fuel by year end 2010. It should signal to the investment community that time to certification is no longer a barrier to investment. CAAFI's R&D and Certification teams are working together to achieve longer term goals of approving sustainable aviation biofuels from sources that may prove to be less capital intensive in the long term. One such process involves fuels from Hydrogenated Plant Oils .

Recognizing that just having a goal of 2013 for sustainable aviation biofuel certification is not nearly enough to realize the CAAFI's goals, its certification and R&D team leaders are working with the Air Force to formulate a joint roadmap co-owned by CAAFI and the Air Force to achieve full certification of new sustainable aviation biofuels identified through the soon to be completed Defense Advance Research Project Agency (DARPA) program..

Ultimately fuel from algae may have the greatest yield of all energy crops, as much as 100 times the yield of current biofuel crops. DARPA and Algae Fuel interests are working closely with industry to launch new projects to head down the road of researching and subsequent commercialization of this promising source.

While Certification of Fuels is a pre-requisite to enabling alternative fuel use in aviation, such alternatives must pass the test of providing significant environmental gains. CAAFI's environmental team has several initiatives launched through the FAA Office of Environment and Energy and the Transportation Research Board's Airport Cooperative Research Program (ACRP) designed to achieve this end. Several key milestones have been met on those projects over the last year. Taken together they place aviation in a position not only of achieving environmental gains but of documenting those gains in ways that are quantifiable. CAAFI is dedicated to making these approaches acceptable to all stakeholders willing to provide input to and consider data based analysis.

The MIT led Partnership for AiR Transportation Noise and Emissions Reduction (PARTNER) Center of Excellence's precedent setting "Well to Wake" environmental life cycle model provides CO2 outputs for over a dozen candidate fuel types. The first phase of the FAA funded project is complete. A second iteration on the study is planned and upgrades to the model to better account for Jet fuel production specifics are planned. It will use "world class" models to assess indirect land use effects in the production of aviation biofuels. Work to date using these models suggest that there are a wide range of fuel options that have the potential to outperform current oil refineries in "well to wake" greenhouse gas production.

Based upon the model and the information that has been provided by those planning the projects in Illinois and Ohio a positive outcome may include mixed coal and biomass to liquid fuel (CBTL) projects now in the planning phase. Those facilities offer the most efficient means to capture CO2 and can use it for enhanced oil recovery. Princeton researchers have indicated that for every barrel of fuel produced, two barrels of oil are obtained from enhanced oil recovery. The mixing of coal and biomass may offer significant improvements in carbon footprint compared to coal to liquid (CTL) plant and oil refineries. Analyses by Princeton and Noblis suggest that carbon neutral synfuel production with as little as 21% biomass content may be possible when by-products are fully utilized.

A key characteristic of the alternative synthetic paraffinic kerosene Alternative fuels currently being brought to market, or in the research phase, is that they are extremely low in sulfur. Sulfur is a precursor to the small soot particles (PM 2.5) that are now identified in EPA issued National Ambient Air Quality Standards (NAAQS). With over half the airports in the U.S. in identified non-attainment areas under the NAAQS use of emerging alternative fuels may be one of the best tools to control growth in PM2.5 exposure that could ultimately limit the growth of commercial aviation in the U.S.

This past week CAAFI introduced a proposal to the Coordinating Research Council (CRC) of the fuels approval agency ASTM (American Society for Materials and Testing) to study the costs and benefits of ultra low sulfur (ULS) fuel in aircraft. Initial studies from MIT show that that control of particle (PM2.5) in jet fuel via ULS fuels (including 100% SPK) could have very sizable public health benefits. Benefits can be double that of the benefits of currently mandated NOX reduction in local airport environments. The study requests the results by the end of 09' for an examination balancing health effect benefits with potential maintenance cost and perhaps to address any airworthiness concerns.

With this data in hand future approvals of pure synthetics post 2009 will be expedited. To ease new fuels introduction to airport markets CAAFI recommended a project to combine these tools via development of an Airport handbook calculation procedure. The Transportation Research Board, Airport Cooperative Research Program (ACRP) initiated that program late last year. The colorfully named project (ACRP project 02-07) is currently in its first phase surveying different sized airports to in part establish needed databases needed to create the handbook and to execute test .cases of handbook use.

One economic key to beneficial introduction of alternative levels of fuels could be the degree to which Ultra low sulfur jet fuels can be used in diesel powered Ground vehicles enabled if 100% SPK fuel can meet the ULS limits now applied to diesel equipment. In airports planning growth the ability to use one, rather than two fuel types provides potential economies. In addition as those airports confronting PM2.5 non attainment could be the best place to introduce synthetics. Identification of those locations could provide an assist to fuel producers in identifying early customers.

Recognizing this, CAAFI interests proposed two new projects to ACRP last week. The first project enhances ACRP 02-07 fidelity by proposing the development of added emissions data for Ground support Equipment from a variety of alternative fuels. The second project proposes a project to prioritize airports that would benefit from the introduction of low PM2.5 fuels non-attainment areas as these areas have been identified in EPA issued National Ambient Air Quality Standards and quantifying what gains are possible.

While action is not complete on FAA Reauthorization it should be noted that the Aviation Committees actions to initiate the CLEEN program, and in the case above to make ACRP permanent and to direct a part of the project to environmental projects, are already serving to make approval of these projects more likely.. Thanks to the Aviation Subcommittee and Transportation Committee for helping to enable these initiatives.

With certified fuel candidates available and needed environmental tools elements in hand the last key needed to launch the aviation alternative fuels in the U.S. are buyers.

The Air Force with its visible goal of 50% use of 50/50 synthetic fuels by 2016 has long been the leaders in the US in this area. As it is with R&D, Certification and Environmental efforts I am pleased to report that the commercial airline industry is now taking important steps to join the Air Force as potential early customers for Synthetic Fuels.

On April 22, Earth Day, the Air Transport Association — a CAAFI sponsor and the representative of U.S Airlines provided its policy on synthetic fuels. A link to that statement is provided in this testimony http://www.airli nes. orq/econom ics/energy/altfuelsp rincip les. htm

The comprehensive statement dedicates the U.S Airlines to work with potential future suppliers to "integrate alternative fuels into their operations". It seeks only alternative fuels that have "a reduced emissions profile relative to traditional fuels on a life cycle basis." The tools that it commits to use to demonstrate that performance are the ones being developed by CAAFI and the FAA Office of Environment and Energy and being used by the Airport Cooperative Research Project as described above.

To ensure that this process is substantive the ATA Energy Council representing the Airlines fuel suppliers and the CAAFI business/economic team now plan to bring suppliers and airline users together in a Department of Commerce hosted meeting in the September /October time period here in Washington to make the needed introductions of a potential new supply base to an aviation industry having unique distribution and quality needs.

These recent developments are extremely important and should signal to potential suppliers and investors that given the availability of competitive business terms for fuels that meet the stated environmental criteria, and a distribution system which is compatible with transport and airport infrastructure, the airline industry is prepared to be a key component in helping to launch new projects.

I am please to advise that as well that the Supplier industry is responding to the message that aviation (both commercial and military) have implemented. In particular the parties planning projects in Illinois (Rentech and American Clean Coal Fuels) and Ohio (BAARD Energy) have indicated in public forums that they are willing to supply up to 70% of facility production to aviation. This amount is far above current Fischer Tropsch facility production volumes which average less than 30% aviation production.

Another indication that suppliers are responding to aviation's call is that increasingly aviation and CAAFI are being invited to address Energy forums regarding our plans. Just next week CAAFI will be the only transportation mode addressing the 8th GTL/XTL summit in London at the invitation of Air Force GTL supplier Shell Aviation.

In November Boeing was a co-sponsor of the first Algae summit held in San Francisco. Aviation interests from several CAAFI sponsors and suppliers were present. It was at this meeting that DARPA announced its multi-million dollar project to enable algae fuels.

With all this excellent progress, largely over the last year, the obvious question is what issues remain as gaps for CAAFI interests including the government agencies to address.

First we must acknowledge is that the current round of projects offers only a small contribution. If all the supply of the potential plants were to be available to aviation they could only supply the 80,000 barrels a day fuel needs of O'Hare airport. Second as we all know the capital needs to flow from to build plant and equipment. Consequently the projects cannot escape the issues of capital formation currently plaguing the investment community. Clearly these issues are well beyond CAAFI's area of influence. We can however ensure that solutions are available when conditions allow more investment.

For elements CAAFI working with the Administration and Congress where we can have influence there are gaps that we do need to address. A partial list includes.

One barrier to sustainable aviation biofuel introduction is the need to produce significant quantities even before the fuel can be certified. For example it is recognized that the engine industry alone (as cited publicly by DARPA) will need up to 250,000 to 500,000 gallons of biofuels from new processes (e.g. Hydrogenated vegetable Oils from their current program) to achieve certification. CAAFI and the airline industry are addressing that issue by adding USDA and the Energy Departments renewables office (EERE) as stakeholders in our process. It is our hope that recognizing the potential for renewable energy suppliers we can find solutions to work this problem. I am certain that interest and support from the Aviation committee with its associates in other Congressional committees can provide assistance.

Technically there a concern with transport of some biofuel types should they become available to airport facilities. CAAFI has rolled concerns of facility compatibility into its definition of a qualified fuel. We know that efforts are underway through such efforts as the Biomass Research and Development Initiative to explore those problems. That said, airport suppliers and pipelines needs to be fully engaged to ensure that our unique needs are addressed. Gasification of biomass while being executed globally suffers in that it has few options and presently has no domestic supplier of needed technology. With options limited we do not have optimal technical solutions. Part of the CAAFI business team's goal is to engage its highly competent systems level- manufacturers in the pursuit of these technologies. Such companies as UTC, GE, Honeywell, and Raytheon in addition to Boeing can potentially help in that quest.

There is a significant concern that there will be adequate amounts and sufficient quantities of water — particularly in western states for the cooling of FT plants and for the production of biomass feed stocks in many areas. Solutions may involve the use of degraded (dirty, salty water). Again Aerospace OEM technologists can help solve the issues which in themselves may be profitable businesses and in the margin help their core aviation business.

Lastly we do need to do ensure that there is a complete and thorough national dialogue and mutual understanding on the quantification of environmental solutions. International harmonization of tools and solutions developed using those tools is a priority for CAAFI's FAA office of Environment and Energy sponsor. For its part CAAFI along with FAA AEE is placing the highest priority on the use of the best indirect land methods, and gasification emissions models specific to aviation in future iterations of its lifecycle analysis tools. This will to ensure the broadest possible buy-in to the outcomes being produced. We need a true international cooperative effort and are increasing our dialogue to include more international partners to approach this goal.

In closing I would like to thank you Mr. Chairman for your support of the CAAFI coalition by providing us with an opportunity to represent our sponsors before the committee.

I would also like to thank publicly the approximate 150, largely volunteer, members of the CAAFI coalition for their contributions to these efforts. For most — including our team leads this is not their "day job". I would like to thank the members of all the stakeholder government agencies from DOE, NASA, USAF, DOD, DARPA, USDA, DOC, and EPA who have crossed organizational stovepipes to cooperate in ways I have not seen in my 41 years in working with government from the vantage of the private sector. I would especially thank Carl Burleson, Dr. Lourdes Maurice and the FAA Office of Energy and the Environment who support were first to recognize the opportunity for the CAAFI coalition and who support my function directly. Without contributions from all of those above the progress reported would not be possible and prospects for the future remote.

That concludes my prepared statement. Thank you.

In: Aviation and Climate Change
Editors: George T. Blumenthal
ISBN: 978-1-60876-757-1

Chapter 4

HEARING ON AVIATION AND THE ENVIRONMENT: EMISSIONS

James K. Coyne

Chairman Costello, Ranking Member Petri, and Members of the Subcommittee:

Thank you for this opportunity to appear before you to discuss our members' efforts to minimize the impact our industry has on the environment.

My name is James K. Coyne, and I am president of the National Air Transportation Association (NATA). NATA, the voice of aviation business, is the public policy group representing the interests of aviation businesses before the Congress, federal agencies and state governments. NATA's over 2,000 member companies own, operate and service aircraft and provide for the needs of the traveling public by offering services and products to aircraft operators and others such as fuel sales, aircraft maintenance, parts sales, storage, rental, airline servicing, flight training, Part 135 on-demand air charter, fractional aircraft program management and scheduled commuter operations in smaller aircraft. NATA members are a vital link in the aviation industry providing services to the general public, airlines, general aviation and the military.

BACKGROUND

Climate change is an important topic and one that includes all of U.S. industry. The aviation industry is one of the fastest growing sectors of the economy, and attention has been focused on carbon dioxide emissions from aircraft. The industry has adopted an environmental agenda that also supports aviation's continued growth. This agenda includes testing alternative fuels to be used in aircraft, participation in carbon offset programs, and encouraging "green airports." It is important to highlight the facts on emissions concerning aviation:

- **Aviation accounts for only 3% of greenhouse gas emissions worldwide,** according to data from the Intergovernmental Panel on Climate Change (IPCC).

- **Aviation gasoline and jet fuel account for 12% of all petroleum products**. According to data from the U.S. Department of Energy, aviation gasoline and jet fuel supply account for 1,624,000 barrels per day compared to 20,588,000 barrels of all petroleum products used per day.

- **Alternative fuels programs exist within aviation.** Virgin fuels, Boeing, and GE Aviation have joined forces to develop an alternative fuel that includes a biofuel blend composed of babassu oil and a mixture of jet fuel and coconut oil. Additionally, the University of North Dakota received a $5 million grant to develop a cold weather sustainable biofuel to be used by the military.

- **New aviation technology is producing cleaner and more efficient aircraft.** Advances in aerodynamics offer more efficient wings and designs with less drag. The Cirrus and Columbia aircraft are examples of more efficient general aviation aircraft. New composite materials are making aircraft lighter in weight, which results in increased fuel efficiency.

NATA CLIMATE INITIATIVE

NATA's members have been very active in addressing ongoing concerns with the general aviation industry's impact on the environment. Late last year, NATA established its

Environmental Committee to develop programs designed to assist member companies in minimizing their impact on the environment including the development of a new Climate Initiative that would provide NATA members with the opportunity to purchase carbon offsets for their aircraft operations. Below is a brief description of the new *NATA Climate Initiative*:

- ***Carbon Offsets:***
  Carbon offsetting involves reducing emissions by investment in projects that save energy, such as investment in technology that allows industry to be more efficient and increasing the generation of renewable energy. NATA is establishing a program to make carbon offsets available to member companies. This includes a 3- to 4-cent per gallon carbon offset based on the Chicago Climate Exchange, the only voluntary, legally binding integrated trading system to reduce emissions of all six major greenhouse gases (GHGs), with offset projects worldwide.

- ***Green Aviation Facilities:***
  NATA encourages its member companies to meet proper environmental compliance standards including a Spill Prevention Control and Countermeasure (SPCC) plan, if fuel is stored above ground in tankers. The association is currently undertaking the development of best management practices that will allow aviation businesses to capitalize on becoming more energy efficient and minimizing their company's impact on the environment. To date, NATA has crafted best management practices for the following topics *(note: each topic has white paper attached at the back of testimony)*:

  - Spill Prevention Control and Countermeasures (SPCC)
  - Hazardous Waste
  - Storm Water
  - Used Batteries
  - Used Oil
  - Used Fluorescent Lamps

PUBLIC RELATIONS CAMPAIGN

NATA is currently developing a public relations campaign to provide the facts about aviation's impact on the environment and what the association's members are doing to ensure the protection of the world's environment.

First, let's review the magnitude of carbon emissions for aviation as compared to other modes of transportation. The U.S. Department of Energy (DOE) has issued a publication titled *Transportation Energy Data Book: Edition 26* (June 1, 2007). In this publication, the DOE provides an annual statistical compendium designed to characterize transportation activity and explore data on other factors that influence transportation energy use. Much of this publication is just the numbers, but an analysis of this data can provide some very interesting insights. In order to obtain a relative comparison of the magnitude of carbon emissions for different modes of transportation, an analysis of fuel consumption is used. Because different types of fuel (gasoline, diesel fuel, jet fuel to name a few) provide different energy values, the data is normalized by looking at the energy use in British Thermal Units (BTUs). This provides a better comparison than actual gallons of fuel consumed. The table below shows the energy use for aviation and several other modes of transportation.

Mode of Transportation	BTUs Used (Trillions) in 2005	Percent of Total
Aviation	**2,477**	**9.0%**
Cars	9,140	33.4%
Light Trucks	8,108	29.6%
Medium/Heavy Trucks	4,577	16.7%
Water	1,366	5.0%
Pipeline	842	3.1%
Rail	657	2.4%

The table shows that aviation accounts for only 9% of the total transportation energy use. Cars and light trucks each use more than three times the energy as the aviation industry, and medium/heavy trucks account for about twice that of aviation. Highway transportation (cars, light trucks, and medium/heavy trucks) combine for almost 80% of the transportation energy and thus contribute a similar level of greenhouse gases. Aviation, on the other hand is a much smaller contributor to energy use and greenhouse gas emissions.

Further analysis within the aviation group shows the relative contribution of commercial aviation versus general aviation. The table below shows this comparison.

Mode of Transportation	BTUs Used (Trillions) in 2005	Percent of Total Transportation
Aviation	**2,477**	**9.0%**
Domestic Carriers	1,861	6.8%
International Carriers	373	1.4%
General Aviation	242	0.9%

General aviation accounts for less than one percent of the total transportation energy use in the United States and its fuel use is about one seventh that of the domestic air carriers.

The analysis of the relative efficiency of fuel use for different modes of transportation includes several additional factors. This analysis includes normalizing the data using the BTU content as did the previous analysis, but also includes the average passengers per vehicle, and provides a conversion into an equivalent miles per gallon (MPGe) based on the BTU content of gasoline (115,000 BTU per gallon). For example, automobiles have an average passenger use of 1.57 passengers per car. Using the DOE data, we find that cars use on average 3,496 BTUs per passenger mile and this corresponds to approximately 33 MPGe. The table below shows several modes of transportation and their corresponding MPGe.

Mode of Transportation	Average Passengers per Vehicle	BTUs per Passenger Mile 2005	MPG Equivalent
Aviation	**90.4**	**3,959**	**29 MPGe**
Cars	1.57	3,496	33 MPGe
Light Trucks	1.72	4,329	27 MPGe
Rail (Commuter)	32.9	2,569	45 MPGe
Rail (Intercity- Amtrak)	17.9	2,760	42 MPGe
Bus (Transit)	8.7	4,318	27 MPGe

This table shows that aviation provides the greatest average passengers per vehicle and the resulting MPGe of 29 is roughly the same as cars and light trucks. With the vast advantage of moving large numbers of people quickly over many miles, the aviation industry is a very efficient mode of transportation.

In summary, this analysis of transportation modes shows that aviation provides a very efficient mode of transportation and compares similarly to typical highway transportation (cars and light trucks) in per passenger equivalent miles per gallon. Additionally, the aviation industry contributes a much smaller percentage (about 9%) of the total energy use and thus contributes a much smaller percentage of greenhouse gas emissions as compared to highway transportation (about 80% for cars, light trucks and medium/heavy trucks).

INDUSTRY ACTIONS

Two prime examples of NATA members being proactive on the environmental front are NetJets Inc. and DayJet Corporation.

NetJets Inc.

On September 13, 2007, NetJets Inc. Chairman & CEO Richard Santulli announced the company's new multifaceted program to address the impact of its flight operations on the environment.

The initiative, which will be expanded in the coming months, includes a focus on offsetting carbon emissions from NetJets' flights, while at the same time it begins to reduce the carbon footprint of NetJets' operations worldwide. It also includes a substantial investment in leading-edge technology research with the goal of creating an ultra-low emission jet fuel.

The following core elements of the initiative are the results of a detailed environmental review process undertaken by NetJets beginning in early 2006:

1. *Improving Energy Efficiency and Reducing Greenhouse Gas Emissions* – NetJets U.S. has established a goal of improving its energy efficiency, cutting waste, and reducing carbon emissions from its internal operations by 10% over the next two years. It has established Director of Environmental Management positions, reporting to the Office of the Chairman, in both the United States and Europe. These senior executives will manage, monitor, and report regularly on NetJets' progress and any ongoing challenges, as well as

helping to identify new opportunities to do more in the coming months and years.

2. *Driving Technological Transformation* – NetJets is investing in cutting-edge research to identify more environmentally friendly aviation technologies through sponsoring The Next Generation Jet Fuel Project at Princeton University with the University of California, Davis to develop an ultra-low emission jet fuel.

3. *Offsetting Unavoidable Impacts* – NetJets is investing in a set of carefully reviewed and closely monitored carbon offset projects that will provide verified greenhouse gas reductions. These projects will allow the company to offset fully the carbon footprint of its internal operations. The offset portfolio will also be available to NetJets Owners so they can offset their flights. Additionally, Marquis Jet Partners will make the NetJets carbon offset portfolio available to Marquis Jet Card Owners.

4. *Leveraging World Class Expertise* – To ensure that NetJets remains a leader on climate issues, the company has established both U.S. and European advisory boards – each consisting of environmental experts who can help NetJets apply best-in-class practices and provide guidance to the company at every step along the way. The U.S. Advisory Board consists of Fred Dryer, Professor of Mechanical and Aerospace Engineering at Princeton University; Ashok Gupta, Director of the Air and Energy Program at the Natural Resources Defense Council; Terry Tamminen, former Secretary of the California Environmental Protection Agency; Bonnie Reiss, Operating Advisor to Pegasus Capital and founder of the Earth Communications Office; and George Favaloro, Managing Partner at Esty Environmental Partners.

NetJets realizes that its responsibilities also extend to the communities in which it operates. With this in mind, it will build on these four areas of immediate commitment by launching a community-based Solar Schools Project. Beginning in California and drawing on Governor Arnold Schwarzenegger's environmental leadership, this innovative public-private partnership will fund the placement of photovoltaic cells on school roofs – while in the process educating school children about environmental issues.

DayJet Corporation

Dayjet Corporation has produced a Very Light Jet Footprint Analysis Concept Paper to ensure that it maximizes its operations by using state-of-the art technology while minimizing its impact on the environment. A description of this concept paper can be reviewed below:

Introduction

DayJet Corporation is a technology-driven company, pioneering the creation of the on- demand, per-seat air, regional transportation market. In spearheading the use of the new generation of very light jets (VLJs) in air carrier service, DayJet is committed to building on the inherent efficiencies and favorable footprint of the Eclipse 500 aircraft, toward the long range goal of sustainable air transportation. DayJet's strategies are informed and motivated by the emerging understandings of the epochal challenges of managing carbon dioxide and other greenhouse gas effects on climate dynamics. DayJet strives to translate the understanding from the climate science community to applications on how the company manages the effects of its aviation activities. DayJet works to apply these understandings to its company strategies for technology, for operations, and for business practices.

DayJet's strategic framework for moving toward a sustainable business in on-demand air transportation includes three major phases of technology implementation over the near term (two to four years), mid-term (four to six years), and longer term (more than six years).

1. **Near Term:** This phase of the strategy focuses on reducing carbon emissions through airspace efficiencies. With the Federal Aviation Administration, DayJet plans the early implementation of certain of the operating capabilities envisioned in the U.S. Joint Planning and Development Office (JPDO) vision for the Next Generation air transportation system (NextGen). This vision underpins transformation of the national airspace toward a performance-based air traffic management system. This system incorporates performance-based navigation, surveillance and communications technologies. The technologies associated with this transformation include Required Navigation Performance (RNP), Area Navigation (RNAV), and Automatic Dependent Surveillance-Broadcast (ADS-B). These technologies let DayJet optimize its flight routes for savings in energy, carbon and noise. When implemented, these optimized routes

will reduce fuel consumption for the company's networked fleet operation by an estimated 15 percent and more. These fuel savings translate directly into emissions reductions. DayJet's business model operates effectively using the second- and third-tier airports serving the nation's smaller suburban, rural and remote communities. These airports are largely underutilized national assets, with virtually none of the congestion and delays issues that would add unacceptable costs to DayJet's operations. At these smaller airports, as a consequence, DayJet is able to operate with greatly reduced ground times from engine start to takeoff and from landing to engine shutdown. The effect is to reduce emissions that would affect local air quality around airports.

2. **Mid Term:** This phase of the strategy focuses on alternative fuels that reduce carbon and other emissions. DayJet supports fuel strategies that enable movement toward carbon neutral practices across the company's industrial sector. Because DayJet's business model calls for approximately a five-year life cycle for its aircraft, the company will be in a position to provide aircraft and engines coming out of its fleet for testing of new fuels.

3. **Longer Term:** This phase of the strategy focuses on being early adopters of advancing aircraft and propulsion systems that do not add carbon and other greenhouse gases to the environment.

Based on this framework, DayJet is undertaking action in five arenas:

- **Engines-** DayJet works with the aircraft and engine manufacturers toward continual improvement in the efficiencies and emissions from flight operations, for the current and future generations of equipment.

- **Fuel-** DayJet collaborates with the university research community on modeling, analysis, and strategies for continual improvement of the footprint for on-demand air transport operations using VLJs and other technologically advanced aircraft. DayJet is engaged in planning dialogue with Embry Riddle Aeronautical University (ERAU), Florida Institute of Technology (FIT), Florida Atlantic University, and the University of Central Florida (UCF) to establish relationship strategies leading to creation and adoption of new technologies.

- **NextGen Airspace Efficiencies-** DayJet partners with other on-demand air transportation service providers to accelerate the early adoption of the operating capabilities envisioned in the U.S. Joint Planning and Development Office (JPDO) Next Generation Air Transportation System (NextGen). NextGen technologies reduce the footprint of DayJet's fleet operations.

- **Neutralizing Carbons-** The DayJet service model is conceived to offer regional business travelers an alternative to highway travel. This means that the aggregation of travelers on a DayJet trip can be carbon-competitive with highway transport alternatives in certain specific cases. In the nearest term, DayJet will implement strategies for airspace efficiencies with the goal of advancing toward carbon-competitive effects of on-demand air travel, relative to highway alternatives.

- **Sustainable Business Practices-** DayJet manages unavoidable carbon emissions through careful assessment of total company operations with the aim of achieving verified greenhouse gas reductions. The goal is to reduce where possible and to otherwise offset the carbon footprint of the company's internal operations.

Tropospheric Flight

The questions regarding potential fuel efficiencies (and therefore carbon sourcing) of on- demand, networked fleet operations of VLJs include the effects of flight in the troposphere versus the stratosphere. The initial assessments by the International Panel on Climate Change (IPCC) regarding the effects of aviation on the environment focused on flight in the stratosphere. These past studies published first in 1999[i] and then reviewed in 2006[ii] summarize the three most important ways that aviation affects the climate:

1. Direct emissions of greenhouse gases specifically CO_2 and water vapor
2. Nitrogen oxide emissions interacting with ozone and methane
3. Contrail-induced cirrus cloud formation

A new generation of aircraft referred to as very light jets (VLJs), and a new generation of on-demand air transportation service business models are now emerging in regions of the United States, as well as in Europe and other continents. These new business models influence the characteristics of fleet operations using these new VLJs. These operations will share common characteristics in flying relatively shorter segments and in making effective use of mid-altitudes that are not extensively used by larger air carrier transports. In particular, because the average flight segment length in these regional transportation services is approximately 300 miles, and because of airspace utilization and traffic flow considerations, the typical flight levels for these fleet operations are below 26,000 feet, well below the stratosphere, and within the troposphere.

Based on these operational realities, we plan to support research focused on the following four topics:

- Comparative greenhouse effect of water vapor emitted by VLJs between 18,000 and 26,000 feet, versus in the stratosphere
- Comparative effect of nitrogen oxide emitted by VLJs between 18,000 and 26,000 feet on atmospheric ozone versus emissions in the stratosphere
- Implications of the effect of NOx-induced depletion of methane for NOx emissions in the troposphere
- Reduced effect of cloud formation due to absence of contrails by VLJs flown in the troposphere
- Effect of lowered propulsive efficiency in the troposphere versus higher propulsive efficiency in the stratosphere on overall carbon dioxide emissions, for VLJs. The results of this analysis will lead to engine design requirements for operating turbofans such as the Geared Turbofan[iii] in the mid altitudes

The state of the art in climate modeling is limited in terms of providing reliable estimates for the specific aviation-induced radiative forcing effects. Even so, perhaps the relative effects of stratospheric and tropospheric flight can be considered. A rule of thumb in climate science estimates that each unit of fuel burned by stratospheric air transport flights has the greenhouse effect of three times that of ground transport[iv]. Further, it is estimated that the radiative forcing from persistent contrails and contrail-induced cirrus clouds (P CC) exceeds the contributions from all other aviation-induced RF combined[v]. However, quantitative analysis is absent regarding the effect of flying in the troposphere specifically by VLJs. We plan to support analysis to provide quantitative information regarding this distinction.

LEGISLATIVE ACTIONS

Clearly Congress will have a vital role in outlining the next steps necessary to mitigate aviation's footprint on the environment. The U.S. House of Representatives' FAA Reauthorization bill, H.R. 2881, provides an outstanding blueprint for the aviation industry to utilize new technologies through the legislation's support of the Next Generation Air Transportation System.

Since last year, NATA members have been strongly encouraged to support H.R. 2881, a bill approved by the U.S. House of Representatives that would provide historic funding levels, nearly $13 billion, for the FAA's Facilities and Equipment (F&E) account that will accelerate implementation of the Next Generation Air Transportation System (NextGen). NextGen is the FAA's national plan to transform the air traffic control system from a ground-based navigation system using radar to a satellite-based system. This legislation will enable the FAA to make needed repairs and upgrades to existing facilities and equipment, and provide for high-priority safety-related systems.

By utilizing new technologies, airspace routes can be better defined, allowing more aircraft and more routes to be determined within the airspace. And most importantly, utilizing new technologies to improve airspace usage, aircraft will be able to fly routes more directly, thereby minimizing noise and the impact on the environment.

The demand for air travel is increasing steadily. The FAA projects that by 2025 the number of domestic enplanements will have doubled to 1,482 million per year. The environmental impact of air travel is increasingly important to consumers, which is why the aviation industry is collaborating to address this important issue.

Thank you for the opportunity to testify, and I will be happy to answer your questions.

End Notes

[i] International Panel on Climate Change Special Report: Aviation and the Global Atmosphere – Summary for Policy Makers, 1999.

[ii] Workshop on the Impacts of Aviation on Climate Change: A Report of Findings and Recommendations. June 7-9, 2006 Cambridge, MA, August 2006.

iii Pratt & Whitney reference TBD

[iv] Noppel, F.; and Singh, R.: Overview of Contrail and Cirrus Cloud Avoidance Technology. Journal of Aircraft, Vol. 44, No. 5, Sept-Oct 2007, pp. 1721-1726.

[v] Sausen, R., Isaksen, I., Grewe, V., Hauglustaine, D., Lee, D., Gunnar, M., Kohler, M. O., Pitari, G., Schumann, U., Stordal, F., and Zerefos, C.: "Aviation Radiative Forcing in 2000: An Update on IPCC (1999)," Meteorologische Zeitschrift; Acta Scientiarium Naturalium Universtatis Normalis Hunanensis, Vol. 14, No. 4, Aug. 2005, pp. 555-561.

In: Aviation and Climate Change
Editors: George T. Blumenthal ISBN: 978-1-60876-757-1

Chapter 5

STATEMENT OF DANIEL K. ELWELL ON AVIATION EMISSIONS

Federal Aviation Administration

Chairman Costello, Congressman Petri, Members of the Subcommittee:

I am pleased to appear before you this morning to address an issue that is central to any discussion of aviation and the environment, aviation emissions. Today I will provide a brief overview of the Federal Aviation Administration's (FAA) activities that help to minimize the environmental impacts associated with aviation emissions, some observations on the current international discussion on emissions trading for aviation, and how Congress can help in moving forward our efforts to address aviation greenhouse gas (GHG) emissions. What should be clear is there is a strong commitment at the very heart of the Next Generation Air Transportation System (NextGen) plan that we have developed--a commitment to provide a systematic, well-informed and performance-based approach to tackling aviation emissions and other environmental issues.

The aviation industry is experiencing record growth globally. It is moving the equivalent of 1/3rd of the world's population each year across the world. Airbus and Boeing have record sales, and two of the fastest growing economies in the world--China and India--are on track to build 100 new airports in the next decade to meet demand.

At the same time, just as aviation is knitting together the world, redefining what opportunity and what neighbor means, concern has grown about its contribution to greenhouse gas emissions and potential impacts on climate change. Aircraft emissions remain a central environmental concern and challenge as they contribute to global climate change and impact the local air quality near airports, and thus could slow the growth of aviation and the benefits it brings to our nation. While we do not have all the answers at this point, what we do have in the NextGen plan is a commitment to provide a systematic, well-informed and performance-based approach to tackling aviation emissions and other environmental issues.

There appears to be a disconnect between perception and performance on aviation emissions, at least in the United States. In some quarters there is a perception that aviation greenhouse gas emissions are growing out of control and that it needs to be reined in by emissions caps and taxes. But consider the facts that we know about performance of the sector and our plans for continued improvement.

Worldwide, aviation represents less than 3% of total man made greenhouse gas emissions. And in the U.S., how have we been doing? The Environmental Protection Agency (EPA) has measured domestic aviation emissions at approximately 3% of GHG emissions. And there is a very positive trend. When you compare today to 2000, U.S. commercial aviation is moving 12% more passengers and 22% more freight while burning less fuel, reducing our carbon output by a million tons. This compares favorably with the U.S. economy overall and aviation has clearly outperfolined passenger vehicles in improving its energy efficiency in the past few decades (see Chart 1).

Now let's give these numbers some context. Consider, for example, the performance of the other major aviation market in the world: the European Union. Between 2000 and 2006, aviation CO2 emissions in the U.S. declined by about 4%. During the same period in Europe, emissions increased by around 30%! In part, this explains our different perceptions of the problem across the Atlantic (see Chart 2).

The fastest means of reducing aviation emissions is to reduce the amount of fuel that is burned. The aviation industry has made and continues to make significant improvements in fuel efficiency. Commercial jet aircraft fuel efficiency has improved 70% [1] over the last 40 years and continues to get better. On a per passenger mile basis, Boeing's new 787 will be as fuel efficient as today's subcompact hybrid car. Also, according to the Air Transport Association (ATA), U.S. commercial airlines have committed to a 30% improvement in fuel efficiency over 2005 by 2025.

Chart 1: U.S. Commercial Aircraft and Automobiles

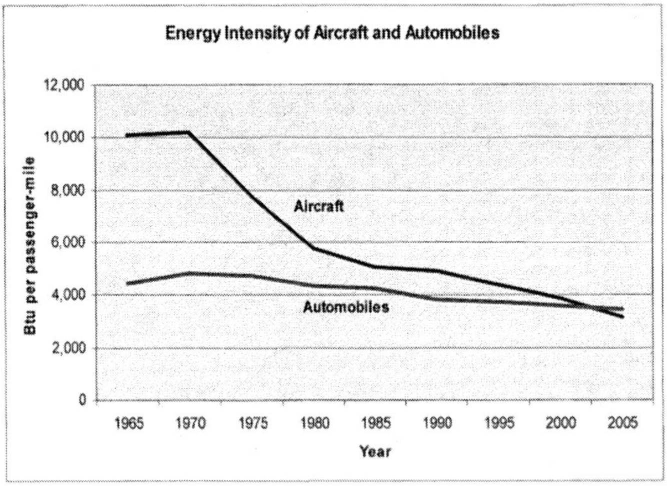

Source: U.S. Department of Transportation, Bureau of Transportation Statistics, *National Transportation Statistics 2007*, Table 4-20: Energy Intensity of Passenger Modes

Chart 2: Differences in GHG Performance

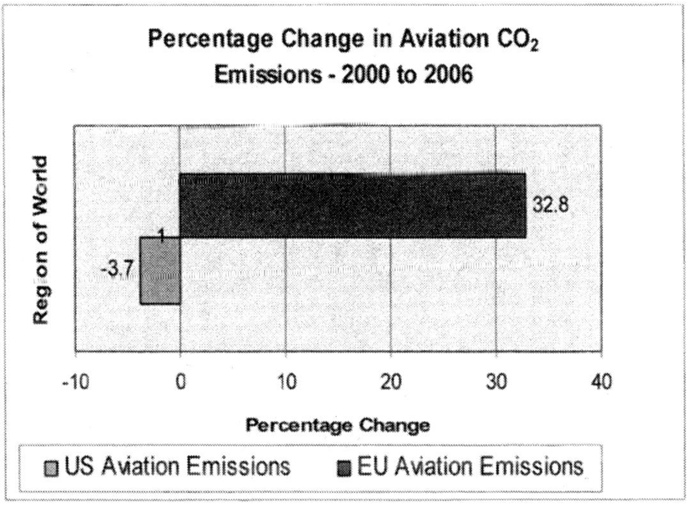

Source: Volpe. (Note: For EU-15)

Chart 3: US Aviation Emissions Growth Down

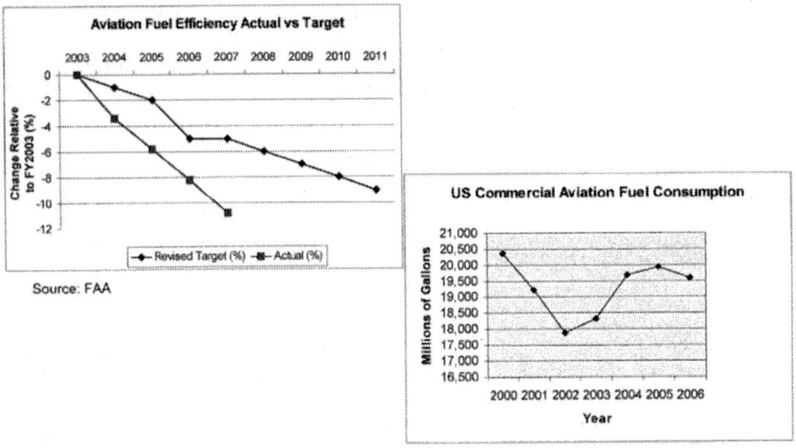

Source: FAA

Source: U.S. DOT, BTS

FAA tracks commercial aviation fuel efficiency and encourages fuel efficiency by U.S. airlines. In just the past four years (2003-2007), U.S. airlines have improved their fuel efficiency 11% (see Chart 3). Since 2000, the restructuring of U.S. airline fleets in the aftermath of September 11th, the rise in fuel costs, utilization of fuel efficient operational procedures, and improvements in air traffic management have all contributed to these savings. With oil now over $100 dollars per barrel and fuel at about a third of operating cost, you can imagine the incentive U.S. airlines have to reduce fuel consumption.

Further, given the weakness of the dollar, the price of fuel for U.S. airlines is about 50% higher than their European counterparts (see Chart 4).

I noted the contribution FAA has made in improving the emissions efficiency of air transport in the United States. Some efforts, like the introduction of Reduced Vertical Separation Minimum (RVSM), have been very successful, saving about 3 million tons of CO_2 annually. RVSM is an International Civil Aviation Organization (ICAO) approved concept that reduces the aircraft separation standard at certain high altitudes, allowing aircraft to safely fly more optimum profiles, gain fuel savings and increase airspace capacity. Other efforts, like the redesign of the Northeast airspace, are more difficult to put in place, but no less important to our overall goal of increasing capacity while minimizing emissions.

Chart 4: Fuel Is Even More Expensive for US Airlines

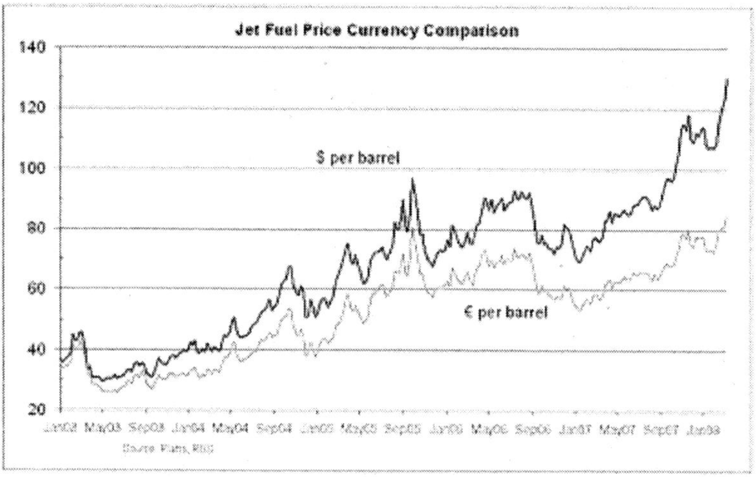

Source: International Air Transport Association (IATA) and Platts

So, the good news is we are starting from a record of exceptional performance historically as we move ahead. So what is our program as we go forward?

Chart 5: Better Understanding of GHG Emissions

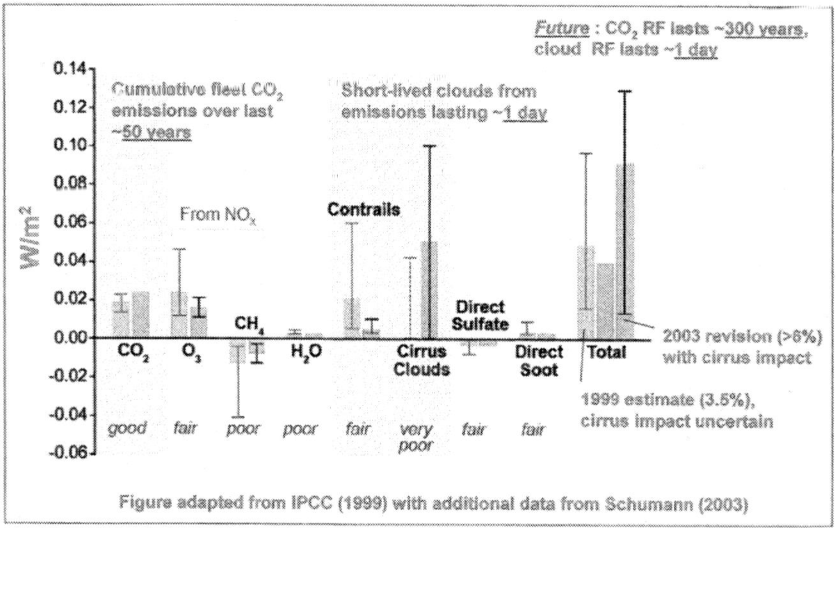

First, we must *improve our scientific understanding* of the impacts of aviation emissions. While CO2's impacts are well known, our understanding of impacts from other emissions—especially at altitude—ranges from fair to poor (see Chart 5). We must ensure that we identify the harmful emissions, accurately measure their impact and design appropriate technologies, or procedures to mitigate or eliminate their effects. This is especially true given the interdependencies that exist—for example, strategies to increase fuel efficiency (and therefore reduce CO2 emissions) can make it more difficult to reduce emissions of nitrogen oxides. As part of our NextGen effort to advance our understanding in this area, we recently launched the Aviation Climate Change Research Initiative (ACCRI) in partnership with the National Aeronautics and Space Administration (NASA) and other agencies. This initiative will help accelerate our scientific understanding to inform policy decisions in this area.

Second, we must *accelerate air traffic management improvements* and efficiencies to reduce fuel burn. Improving energy efficiency has the dual benefit of improving both environmental and operational performance of the aviation sector. As I said before, we have saved millions of tons of carbon emissions over the past couple of years by putting RVSM in place. We are accelerating implementation of other enhanced air traffic control navigation and other procedures to further improve the fuel efficiency of the system. Through the use of Required Area Navigation (RNAV) and Required Navigation Performance (RNP) technology, aircraft will be able to use descent procedures that bum less fuel and result in quieter operations. In addition, satellite-based air traffic control paired with Automatic Dependent Surveillance-Broadcast (ADS-B) technology on aircraft allow for safer but closer separations between aircraft and more direct routing, which will improve fuel efficiency and also reduce carbon dioxide emissions. In essence, NextGen itself will improve environmental performance. We are already achieving early gains at a test program at Dallas-Fort Worth International Airport, where American Airlines' use of NextGen-related procedures is reducing carbon dioxide emissions by levels equivalent to removing 15,000 cars from the road for a year.

A good example of emissions reductions from aviation operational improvements is Continuous Descent Arrival or CDA. CDA allows an airplane to fly a continuous descent path to land at an airport, rather than the traditional "step downs" or intermediate level flight operations. The airplane initiates descent from a high altitude in a near "idle" engine (low power) condition until reaching a stabilization point prior to touch down on the runway. Trials in Louisville, KY have shown a fuel savings (and thus GHG savings) averaging about 12% for the arrival portion of the flight. And testing at Atlanta Hartsfield International Airport of continuous descent arrivals shows savings of 1,300 pounds of carbon dioxide for each and every flight.

CDA is one of those win-win strategies, having environmental and operational benefits that can reduce noise, emissions, and fuel burn, as well as flight time. The cumulative impact of measures like this throughout the system can have a real impact. As additional advanced aircraft and air navigation procedures planned for the NextGen system are developed and deployed, we will see an even greater reduction in greenhouse gas emissions impacts from aviation.

Third, we must *hasten the development of promising environmental improvements in aircraft technology*. This builds upon the fact that the vast majority of improvements in environmental performance over the last three decades have come from enhancements in engine and airframe design. Both the House and Senate have included a number of our environmental proposals in their pending aviation reauthorization bills (H.R. 2881 and S. 1300) including a proposal to create a research consortium, to be called CLEEN--Continuous, Low Energy, Emissions, and Noise--focused on accelerating the maturation of lower energy, emissions and noise technology for aircraft. While action on that legislation is not completed, we already have in place a cooperative working relationship with NASA and broad participation of outside stakeholders through our research advisory committee, the Partnership for AIR Transportation Noise and Emissions Reduction (PARTNER) Center of Excellence advisory board, and our NextGen Environmental Working Group.

Fourth, it is imperative to *explore the potential of alternative fuels for aviation*--fuels that could have benefits for energy security as well as emissions performance, depending on the fuel's lifecycle greenhouse gas emissions profile. The FAA is a major partner in the Commercial Aviation Alternative Fuels Initiative, or CAAFI. CAAFI's participants, which include a cross-section of airlines, manufacturers, airports, fuel producers, federal agencies and international players, are implementing a road-map to explore the use of alternative fuels for commercial aviation. Let me emphasize this is not "pie in the sky". CAAFI participants have already used coal-to-liquid and gas-to-liquid fuels in jets, and most recently completed a bio-fuels flight demonstration. We are keenly aware production processes could increase the overall carbon footprint, so CAAFI is doing careful life cycle carbon emissions analyses and focusing on approaches that will lead to overall reductions.

Data indicate that low sulfur synthetic and bio-based fuels promise significant health promise reduced carbon emissions. To begin to measure these, FAA sponsored a life- cycle analysis of the "well to wake" greenhouse gas emissions of multiple alternative fuels in a study due this spring that addresses the feasibility of alternative fuels for aviation.

Finally, a variety of *market-based measures may offer assistance in managing aviation emissions growth.* Approaches using tax incentives, emissions trading or carbon offsets may all have a role to play, though each can pose challenges in design and implementation. Consider carbon offsetting. This is a scheme which allows airline passengers to pay for carbon reductions accomplished somewhere else to compensate for the emissions generated by the aircraft flight they took. While offered by several airlines, a number of questions have arisen related to calculations of carbon emissions (calculations of the same flight can produce carbon numbers that vary by a factor of three) and how the funds collected are spent. More recently in the U.S. we are looking for market-based measures to increase utilization of congested airspace, so that we can simultaneously increase efficiency and drive down emissions per passenger.

With respect to emissions trading, the U.S. participated in the development of emissions trading guidance for aviation under the auspices of ICAO, the United Nations standard setting organization of international aviation. The U.S. and the rest of the world, except for Europe, agreed on this guidance last September for countries that decide to employ emissions trading for international aviation. The overwhelming majority of countries--developed and developing--all agreed emissions trading should only be applied to another country's airlines on the basis of agreement between States. European countries refused to join consensus, as their proposed legislation would force international airlines into their emissions trading system without the consent of governments

The U.S. has significant concerns about the European Union (EU) legislation that is currently being developed to place aviation into their emissions trading system. On top of the legal issues with respect to the Chicago Convention and our air services agreements, recent discussions with EU officials made clear that adoption of emissions trading for aviation has become an end in itself, rather than improving environmental performance. The facts that U.S. airlines pay substantially more for their fuel than their European competitors, that the U.S. has a domestic fuel tax unlike their EU competitors, and that U.S. airlines have actually reduced their emissions unlike the substantial growth from EU airlines, were dismissed.

As ICAO recognized in its work, an emissions trading system is only one approach and it remains the decision of a State whether to employ such a measure. Market based measures can reduce emissions at lower costs. However, the price of fuel already provides both airlines and manufacturers strong market incentives to reduce fuel consumption. Between 1985 and 2004, aviation outperformed every other transport mode in reducing its emission intensity (see Chart 6). Between 2000 and 2006, the price of fuel more than doubled. Consequently, U.S. commercial carriers bought 750 million fewer gallons in 2006 than they purchased in 2000 even while carrying twelve percent more passengers and 22 percent more cargo. This lends support to the 2001 finding of ICAO's Committee on Aviation Environmental Protection (CAEP) that the price of fuel obviates the need for CO2 emissions standards for aviation.

Environmental advances in the aviation sector historically have been most helped by positive economic measures that further stimulate research and innovation in the industry's fleets. As the record on aircraft noise and fuel efficiency demonstrates, implementation of new technology and operational procedures have been remarkable tools for limiting and reducing aviation environmental impacts.

Chart 6: Energy Intensity Trends by Transportation Mode

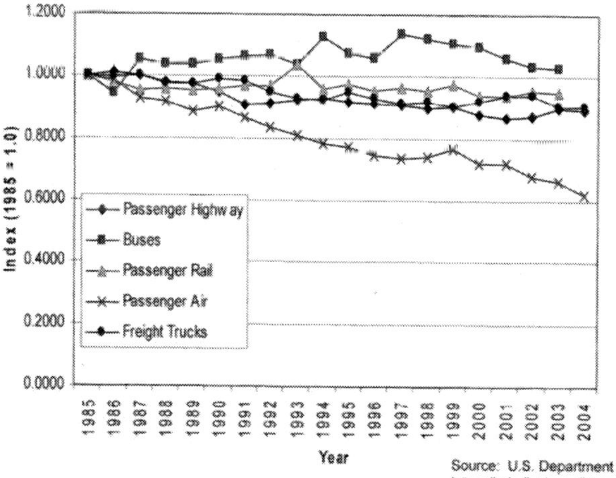

Source: U.S. Department of Energy, U.S. Energy Intensity indicators. (intensityindicators.pnl.gov)

As a recent Congressional Budget Office report (February 2008) highlighted, use of emissions trading as a market-measure to reduce emissions poses a number of issues. FAA remains concerned that such issues become more complex when dealing with aircraft that operate internationally. A poorly designed and implemented emissions trading system could actually hamper the ability of aviation to become cleaner and quieter.

We believe ICAO must continue to exercise global leadership to achieve aviation growth in an environmentally responsible fashion. ICAO offers the best forum to find the harmonized approaches we need for a global industry like aviation. It allows the proper balance of collaboration and State sovereignty. We are committed to supporting that effort. In February, I represented the U.S. at the first meeting of the fifteen-nation Group on International Aviation and Climate Change (GIACC). This high-level group was conceived during last year's ICAO Assembly and is developing an international plan to address international aviation greenhouse gas emissions. Our hope is to take the approach I have outlined here--a balanced approach derived from the recognition that operational and technological environmental performance improvements, coupled with market measures where necessary, can form the basis to derive data-driven, challenging, aspirational goals for the international community in reducing the growth of aviation's greenhouse gas emissions impacts. At the GIACC, we ultimately seek an effective, globally devised strategy, collaboratively entered into.

In addition to FAA's work at ICAO we are pursuing partnerships with other authorities and the international industry in a number of highly technical system areas to advance improvements in aviation's environmental performance. For example, last year the FAA and European Commission (27 countries) announced the Atlantic Interoperability Initiative to Reduce Emissions, or AIRE. The AIRE initiative is targeted to undertake demonstrations in both the U.S. and Europe to accelerate the ability of airlines and air navigation authorities to employ enhanced air traffic procedures that reduce aviation's emissions and noise footprint on either side of the Atlantic. We (the U.S., Australia and New Zealand) also just launched a similar initiative in the Pacific—ASPIRE–or the Asia and South Pacific Initiative to Reduce Emissions.

Aviation has succeeded in its first century because it has constantly met the challenge of innovation and record setting – flying faster, cleaner, quieter and safer. In doing so, aviation has transformed the world. Any fair reading of history will show that until now, aviation has done an exceptional job in improving its environmental performance. But to be blunt, the issue is not past performance, but what we are doing for the future.

In closing, it is clear today that aircraft emissions impact the climate, are an issue of both domestic and international concern and remain a potential constraint on the future growth of aviation. It is also evident we have no "silver bullets." What we do have is an approach to reduce aviation greenhouse gas emissions in a growing NextGen system. We have already initiated a number of endeavors – "silver buckshot" if you will – that will help get us there. We need the help of Congress. We have outlined a significant set of initiatives underway to address aviation emissions. We have proposals before Congress in FAA's reauthorization proposal that, if authorized and funded, would accelerate all these efforts.

Success will require partnership and shared responsibilities among many stakeholders— with air carriers operating cleaner and quieter aircraft; airframe and engine manufacturers improving efficiency of their products; air traffic management facilitating environmentally-friendly procedures consistent with safe and efficient operation; alternative fuel producers scaling up environmentally sound fuel production; airports investing in cleaner infrastructure; and federal programs and investments supporting the necessary technology and operational improvements. The FAA is committed to working with all stakeholders to find the right balance to manage capacity growth while addressing aviation emissions.

Mr. Chaizivan, that completes my prepared statement. I would be happy to answer any questions you or Members of the Committee may have.

End Notes

[1] Intergovernmental Panel on Climate Change (IPCC) special report entitled, *Aviation and the Global Atmosphere,* 1999.

In: Aviation and Climate Change
Editors: George T. Blumenthal

ISBN: 978-1-60876-757-1

Chapter 6

WRITTEN TESTIMONY OF DR. DAVID W. FAHEY ON AVIATION AND THE ENVIRONMENT: EMISSIONS

United States Department of Commerce

INTRODUCTION

Good afternoon, Mr. Chairman and members of the Committee. I am David Fahey, research physicist in the National Oceanic and Atmospheric Administration's (NOAA's) Office of Oceanic and Atmospheric Research. NOAA's mission is to understand and predict changes in the Earth's environment and conserve and manage coastal and marine resources to meet our nation's economic, social, and environmental needs. The Office of Oceanic and Atmospheric Research conducts and sponsors the scientific research, environmental studies, and technology development needed to improve NOAA's operations and applications, and broaden our understanding of the Earth's atmosphere and oceans.

I work at NOAA's Earth Systems Research Laboratory in Boulder, Colorado where I conduct research related to the chemical and particle composition of the atmosphere and work with policymakers to describe and evaluate the science of ozone depletion and climate change. As part of my research over the past 10 years, I have been investigating the role emissions from aviation operations play in climate change. I was a coordinating lead author of the chapter on aviation-produced particles and cloudiness in the Intergovernmental Panel on Climate Change's (IPCC) Special Report on *Aviation and the Global Atmosphere* published in 1999. More recently I was a lead author of a chapter of the IPCC 4th climate science assessment released in 2007 that included an evaluation of the influence of global aviation on climate.

Thank you for inviting me to discuss the impact of aviation emissions on climate. Today, I will provide an introduction to this hearing by presenting some aspects of our basic understanding of the role of global aviation in climate change, the key uncertainties in that understanding, and outstanding gaps in our knowledge.

Aviation is one component of human activities that contribute to climate change. Human activities contribute to climate change by altering the natural amounts of greenhouse gases, small particles, or cloudiness in Earth's atmosphere. Greenhouse gases, particles, and clouds affect climate by influencing the balance between incoming solar radiation from the sun and the outgoing infrared (thermal) radiation. This radiation balance is what controls Earth's temperatures and other climate features. When humans change the specific amounts or certain properties of greenhouse gases, particles, and clouds in the atmosphere, they change the radiative balance and create what scientists call a *radiative forcing* of the climate system. Radiative forcing, or RF as it is often designated, is the widely accepted measure of how hard the climate system is being *pushed* away from its natural state. This push is also known as climate forcing, and can be caused by natural mechanisms or through manmade pollution, such as the emissions of greenhouse gases. We know that if the climate is forced too hard, the climate state will change, altering basic climate parameters such as temperatures and precipitation. An important goal for climate scientists is to quantify climate forcing from all human activities and estimate how and when our climate system might respond to the total forcing. It is in this context that the contribution of aviation to climate change is most appropriately viewed and evaluated.

The radiative forcing contribution of aviation has been the focus of international scientific interest for some time. In 1999, the IPCC released its special report titled *Aviation and the Global Atmosphere*, which comprehensively addressed, for the first time, the processes by which aviation leads to radiative forcing. U.S. scientists, including myself, played leading roles in this effort. This report and subsequent refinements in the IPCC 4th scientific assessment of climate change released in 2007 are the preferred basis for any scientific discussion of aviation's contribution. The uncertainties and knowledge gaps identified in the 1999 report, while reduced in some respects in the intervening years, remain an important limitation in our understanding of the issue.

WHAT ARE THE ASPECTS OF AVIATION OPERATIONS THAT LEAD TO CLIMATE FORCING (OR CHANGE)?

There are three aspects of aviation that are of key importance in understanding and defining its role in climate change.

- *Aviation emits gases and particles into the atmosphere.* Global aviation burns fossil fuel, primarily in jet engines, to propel and operate a wide variety of aircraft. The combustion of aviation fuel (kerosene) produces a variety of gases and particles in the exhaust. Primary among them are the gases carbon dioxide (CO2), nitrogen oxides (NOx) and water vapor, and particles composed of soot and sulfate (e.g., H2SO4). All of these components contribute to climate forcing.

- *Some gases and particles emitted by aviation at cruise altitudes have enhanced roles* in climate change processes. A large fraction of aviation fuel is burned at cruise altitudes (greater than 25,000 ft). At these altitudes, the removal of some gases and particles from the atmosphere is slower, thereby allowing these components to accumulate and stay in the atmosphere longer and have greater effect than they otherwise would if the emissions had occurred near Earth's surface.

- *Aviation operations often increase cloudiness along and near aircraft flight tracks.* Under certain atmospheric conditions, jet aircraft produce contrails (condensation trails), which are a form of cirrus cloud, in the engine exhaust plume. These contrails can sometimes persist and spread for hours to days, depending on conditions (see Figure 1). Since clouds are an important aspect of Earth's radiation balance, increasing cloudiness is a component of climate change.

Figure 1. Persistent contrails spreading to form induced cirrus as viewed from the ground from a location in suburban Washington, DC, USA (EPA, USA Environmental Protection Agency (EPA), Aircraft Contrails Factsheet, http://www.epa.gov/otaq/regs/nonroad/ aviation/contrails.pdf)

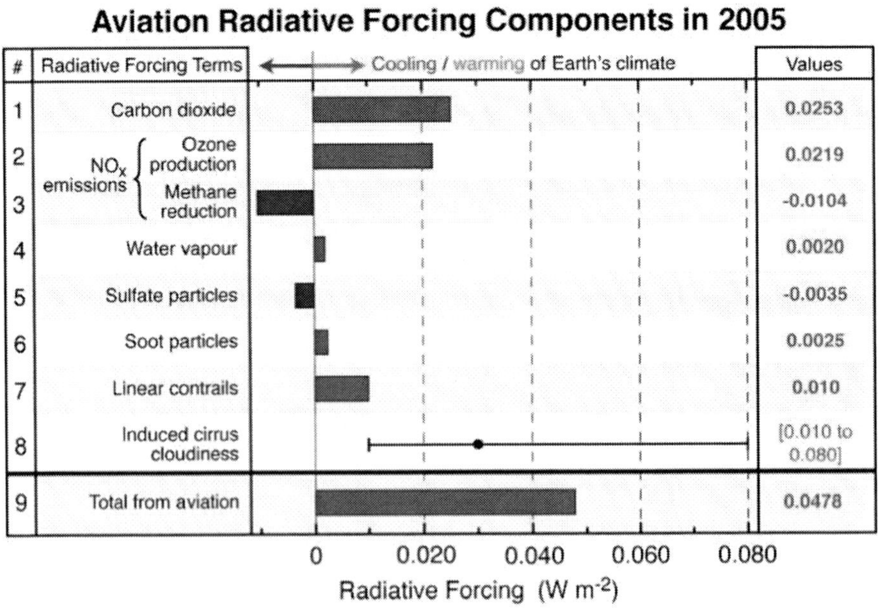

Chart 1. Globally averaged radiative forcing estimates for the affects of aviation on climate. The results account for all aviation operations through 2005. The black line for induced cirrus cloudiness represents a range for the best estimate. (Adapted from IPCC, 2007).

The IPCC reports in 1999 and 2007 quantified many of the climate effects of aviation emissions and cloudiness. Chart 1 graphically illustrates the radiative forcing from each of the eight principal components associated with aviation as reported by IPCC. The quantitative values are radiative forcings given in units of *watts per meter squared* (Wm^{-2}), which is the accepted nomenclature within IPCC. *Positive* radiative forcing values shown in Chart 1 lead to a warming of Earth's climate while *negative* forcings lead to a cooling. Larger forcings are expected to cause larger climate responses. Each bar in the chart represents the *best estimate* of scientists using available data and atmospheric models. I will briefly summarize each component for the committee.

1. *Carbon dioxide (CO2).* Carbon dioxide, a well-known greenhouse gas, is the largest emission (by mass) of aviation. It is a direct product of the combustion of aviation fuel, which primarily contains elemental carbon, hydrogen, and sulfur. The radiative forcing from carbon dioxide is positive (warming) and is associated with a small uncertainty. The climate role of carbon dioxide emissions from aviation is no greater or less than the same amount of surface emissions of carbon dioxide because of its long lifetime in the atmosphere.

2-3. *Nitrogen oxides (*NO_x*).* Nitrogen oxides are a byproduct of high-temperature combustion of aviation and other fuels. Most nitrogen oxides are not greenhouse gases but influence climate indirectly by causing changes in other greenhouse gases; namely ozone (O3) and methane (CH4). In chemical processes that occur in the sunlit atmosphere, nitrogen oxides lead to ozone formation and methane reductions. Methane is emitted in other human activities. These effects have opposite climate effects with ozone increases causing a positive forcing (warming) and methane decreases causing a negative forcing (cooling).

4. *Water vapor.* Water vapor (H2O), a potent greenhouse gas, is a direct product of the combustion of aviation fuel. The highest altitude aircraft emissions occur in the lower stratosphere, where water vapor abundances are low. The accumulation of aviation water emissions in this region leads to a small positive radiative forcing (warming). The accumulation of water emitted at lower cruise altitudes (troposphere) has a negligible climate effect because the natural abundance of tropospheric water vapor is far larger.

5-6. *Sulfate and soot particles.* Sulfate and soot particles are emission products of aviation. Sulfate particles derive from the sulfur content of the fuel. Soot particles are a byproduct of incomplete combustion. The direct effects of the accumulation of these particles are opposite: sulfate particles reflect sunlight causing a small negative forcing (cooling) while soot particles absorb sunlight causing a small positive forcing (warming).

7. *Persistent contrails.* Persistent (linear) contrails form in the jet engine exhaust plume when atmospheric humidity conditions are favorable for ice cloud formation (*i.e.,* supersaturation). Contrails are ice clouds that are formed primarily from atmospheric water vapor but whose formation is triggered by emitted water vapor. Natural and contrail cirrus clouds can both cool and warm the atmosphere depending on specific cloud properties. Contrails from the current aviation fleet are estimated to cause a net positive forcing (net warming).

8. *Induced cloudiness.* Induced cloudiness, also known as contrail cirrus, is defined as cirrus cloudiness that spreads or evolves from persistent contrails as they lose their characteristic linear shape. Induced cloudiness represents cloudiness that otherwise would not have occurred in the atmosphere. As with contrails, induced cloudiness has a net positive forcing (warming) of climate. The forcing for induced cloudiness is shown as a range in Chart 2 because no best estimate value is available.

Total aviation radiative forcing: The best estimate of the total aviation radiative forcing is positive, leading to a warming influence on climate. The total is a sum over the individual components except for induced cloudiness, which is excluded because it lacks a best estimate. Thus, the total shown is an underestimate of the actual total contribution from aviation. The contributions from carbon dioxide and ozone are the largest terms in the total. The total aviation radiative forcing in Chart 1 can be compared to the total forcing from other human activities and the natural forcing from changes in solar output as displayed in Chart 2. The comparison shows that aviation in 2005, excluding induced cloudiness, represents approximately 3 percent of that from all human activities since the start of the industrial era.

An important aspect of the contributions of aviation to climate forcing is the *lifetimes* of the individual contributions. The lifetime refers to how long a particular effect persists in the atmosphere. All of the aviation components except carbon dioxide have short lifetimes. Carbon dioxide emitted into the atmosphere has a lifetime of 100-1000 years. Thus, a large fraction of the carbon dioxide emitted into the atmosphere from aviation in the year 2000 will still be contributing to climate forcing in 2100. This is true for carbon dioxide emitted in all other human activities. In contrast, the effects of other emissions and clouds from aviation have short lifetimes (less than 1 year). Thus, the non-carbon-dioxide effects of aviation in 2000, for example, will represent a negligible contribution to climate forcing well before 2100. Another consequence is that in a future scenario in which aviation operations remain constant, the contribution of carbon dioxide forcing relative to all other aviation forcings would continually increase. The contrast in lifetimes for the different effects is an important aspect of the longterm influence of aviation on climate.

What Are the Uncertainties in Evaluating the Impact of Aviation Operations on Climate Forcing (or Change)?

The climate effects of aviation are the result of complex interactions involving emissions and clouds in the atmosphere. As a consequence, the effects must be calculated with computer models of the global atmosphere. The processes contained in models are an attempt to represent comprehensively the chemical transformations among atmospheric trace gases, the formation and removal of particles, atmospheric air motions that transport gases and particles throughout the atmosphere, the regional and seasonal responses of the radiative forcings, and the response of the climate system to the applied forcings. The resulting model estimates are associated with varying uncertainties because our knowledge of the processes and our ability to represent them quantitatively in models is imperfect. In this presentation, I address only the uncertainties associated with the forcings.

The largest term in Chart 2, that of carbon dioxide, has one of the smallest relative uncertainties, in part, because carbon dioxide is gas with a long atmospheric lifetime and is well studied as a principal greenhouse gas associated with many non-aviation human activities.

The two radiative forcing components from nitrogen oxide emissions have large uncertainties because complex, chemical processes link nitrogen oxides to ozone and methane changes, and the natural abundances and variabilities of ozone and methane are generally larger than the calculated changes.

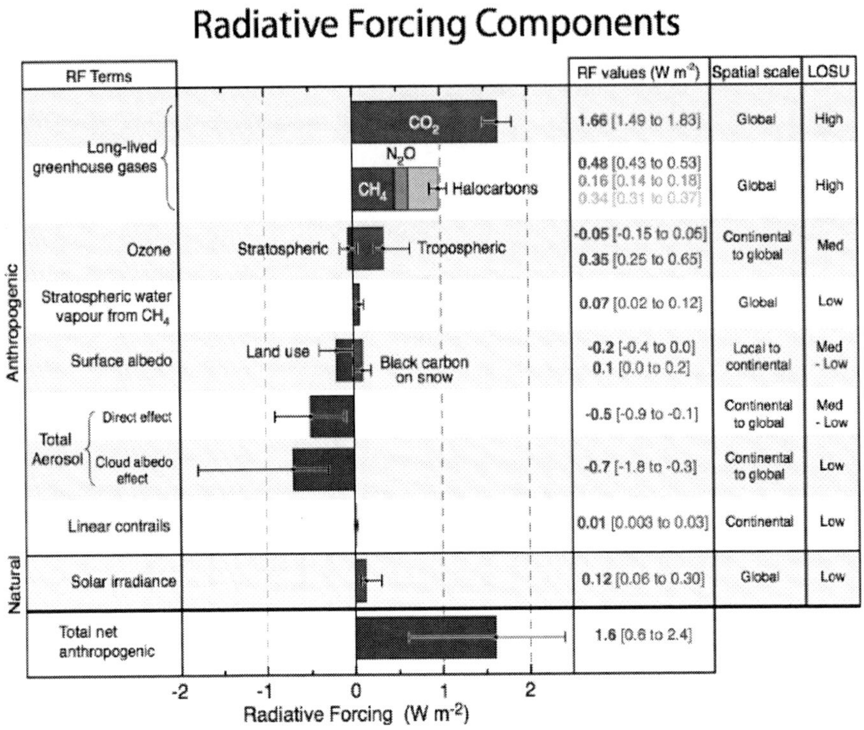

Chart 2. Globally averaged radiative forcing estimates and uncertainties for human-induced increases in principal greenhouse gases and for other important agents and mechanisms between pre-industrial times and the present (2005). Also shown are the typical geographical extent (spatial scale) of the forcing and the assessed level of scientific understanding (LOSU). (IPCC Summary for Policymakers, 2007; Figure SPM2)

Contrail radiative forcing also has a large uncertainty. Contrails, despite being a visible component of climate change, are difficult to assess on the global scale required for aviation. Contrail formation on any flight is critically dependent on the atmospheric humidity conditions along the flight track. Global models have difficulty predicting atmospheric humidity along air traffic routes with the needed precision and accuracy.

The uncertainties associated with the remaining components; namely water vapor and sulfate and soot particles, are relatively small in comparison with other uncertainties. The radiative forcings associated with these components is also small relative to the carbon dioxide component.

WHAT ARE THE GAPS IN OUR KNOWLEDGE ON CLIMATE FORCING FROM AVIATION?

Significant gaps exist in our knowledge of the effect of aviation on current and future climate. Two gaps are worth noting in this presentation:

Estimate of induced-cirrus cloudiness. Currently no best estimate is available for the climate forcing of induced cloudiness from aviation. Many observations have shown the existence of induced cirrus. Chart 1 shows a common example of spreading contrails forming induced cirrus. However, scientists lack the necessary observations and framework to detect and quantify induced cirrus for global aviation. Principal difficulties are distinguishing induced cirrus from background cirrus that forms in the same region and establishing the radiative forcing properties of induced cirrus. Without a best estimate for induced-cirrus, the aviation impact on climate will be incomplete and likely underestimated.

Estimate of the role of aviation particles in background cloud formation. Aviation particles containing sulfate, soot, and unburned hydrocarbons have the potential to alter the formation and properties of clouds in the atmosphere. Studies show that aviation likely increases the number of soot particles, for example, in the upper atmosphere, particularly near flight corridors. The presence of these particles in regions of cloud formation could potentially alter the formation rates of clouds or their radiative properties. Studies have confirmed these effects in the lower atmosphere for other sources of particles from natural and human activities, but estimates are uncertain. Studies have not been undertaken to demonstrate these effects for aviation particles.

SUMMARY

Global aviation emits gases and particles into the atmosphere that affect Earth's climate. The gases include carbon dioxide, a principal greenhouse gas. The emissions occur primarily at cruise altitudes, which increases their potential to cause climate effects. In addition, aviation indirectly increases cloudiness through the formation and spreading of persistent contrails. The net result of aviation emissions and cirrus cloudiness is a positive radiative forcing which leads to a warming of climate. Global atmospheric models are required to evaluate and quantify the separate effects of aviation on climate. Model calculations for aviation are associated with notable uncertainties, particularly for contrails and the indirect effects of NO_x emissions. Identifiable gaps in our understanding of aviation effects are the lack of best estimates for the forcing of induced cirrus formed by spreading contrails and for the possible effect of aviation particles on background cirrus formation and properties. The lack of a best estimate for induced cirrus leads to an underestimate of aviation climate forcing. The lifetimes of the non-carbondioxide effects of aviation are dramatically shorter than the effect of carbon dioxide. The contrast in lifetimes for the different effects is an important aspect of the long-term influence of aviation on climate.

REFERENCES

[1] IPCC, (1999). *Aviation and the Global Atmosphere: A Special Report of IPCC Working Groups I and III* [Penner, J.E., et al. (eds.)]. Cambridge University Press, Cambridge, United Kingdom and New York, NY, USA, 373.

[2] IPCC, (2007). Changes in atmospheric constituents and in radiative forcing, in Climate Change 2007: The Physical Science Basis - Contribution of Working Group I to the *Fourth Assessment Report of the Intergovernmental Panel on Climate Change*, edited by S. Solomon et al., p. 129, Cambridge Univ. Press, New York.

In: Aviation and Climate Change
Editors: George T. Blumenthal

ISBN: 978-1-60876-757-1

Chapter 7

AVIATION AND THE ENVIRONMENT: EMISSIONS AND THE COMMERCIAL AIRLINES' CLIMATE CHANGE COMMITMENT

James C. May

Thank you, Mr. Chairman. ATA airline members transport more than 90 percent of all U.S. airline passenger and cargo traffic.[1] Our airlines take their role in controlling emissions very seriously. Recently, there has been a great deal of focus in Congress on greenhouse gas (GHG) emissions in particular, and how this nation might achieve reductions in these emissions while maintaining economic stability and enhancing energy independence. Commercial aviation has a vital role to play in this regard and I appreciate the opportunity to appear before you today to discuss what we are doing to tackle this important issue.

INTRODUCTION AND OVERVIEW

For generations, flying has contributed to a better quality of life in America. Commercial aviation has been essential to the growth of our economy, yielded breakthrough technologies, brought people together and transported critical cargo – all while achieving an exceptional environmental track record. Today's airplanes are not just smarter – they are quieter, cleaner and use less fuel than ever before – but we also fly them smarter. That's why our industry represents just two percent of all GHG emissions in the United States, while driving three times more economic activity. But we are not stopping there. The initiatives that we are undertaking to further address GHG emissions are designed to responsibly and effectively limit our fuel consumption and GHG contribution while allowing commercial aviation to continue to serve as a key contributor to the U.S. economy. At the same time, while our industry represents well under one percent of the nation's oxides of nitrogen (NO_x) local emissions inventory, we continue to support technological and standards developments to further reduce NO_x output from aircraft. I want to emphasize three points that are essential to moving our emissions-reducing efforts forward:

First, ***commercial airlines are extremely GHG efficient.*** For the past several decades, commercial airlines have dramatically improved our GHG efficiency by investing billions in fuel-saving aircraft and engines and innovative technologies like winglets and cutting-edge route optimization software. Fuel is our largest cost center, creating an economic imperative to maintain our record of continuously improving GHG efficiency. And while commercial aviation accounts for only two percent of domestic man-made GHG emissions, we shepherd this to good use, driving a far larger percentage of economic activity, not only directly, but also indirectly, as a necessary element in the airport and tourism sectors and in all business sectors that rely on the rapid delivery of goods and human resources.

Second, ***ATA airlines are proactively committed to further limiting their emissions footprint*** through a set of measures that will simultaneously address climate change, energy independence and local emissions impacts while preserving economic stability and the opportunity to grow. At the core of these measures is the ATA carriers' commitment to an additional 30 percent fuel efficiency improvement by 2025 – improvement that only comes from the airlines' investment in new aircraft, new aircraft engines, navigation aids and enhanced operational procedures. In addition, we are dedicating ourselves to developing commercially viable, environmentally friendly alternative jet fuel, which could be a game- changer in terms of aviation's output of GHGs and NO_x. Moreover, we are central stakeholders in partnering efforts to modernize the outdated air traffic management (ATM) system and to reinvigorate research and development in aviation environmental technology, both of which can bring extensive additional emissions reductions.

Third, ***there is a critical role for the federal government to play***, not *for* the industry and hopefully not *against* the industry, but, rather, *with* it. While the ATA airlines' 30 percent fuel efficiency improvement target will be met purely through the airlines' investments and operating initiatives, the other measures in the package require a significant measure of congressional support. For example, although we are working with the Federal Aviation Administration (FAA) on plans to replace the antiquated ATM system – an upgrade that promises to bring 10-15 percent emissions improvement *on top* of the ATA commitment – congressional approval is needed before significant progress can be made in implementing this system. Further, the commercial airlines cannot stimulate the development of environmentally friendly alternative jet fuel and aircraft environmental technology on our own. Congressional support and funding and other incentives are vital to these research-intensive initiatives.

Just as we ask Congress to continue to work with us, we also urge Congress to calibrate any climate change-related legislation so it does not work against our efforts. To have the resources to continue our fuel efficiency and other advances, we must have the capital to invest in newer aircraft and other emissions-reducing measures. Punitive economic measures that siphon funds out of our industry would severely threaten that capability, as would unilateral efforts that do not take the international nature of aviation into account. A vibrant, competitive and growing aviation sector is a key part of the solution –not an impediment to ensuring a future where a strong economy, freedom from foreign oil and cleaner air are the order of the day.

Commercial Aviation is Extremely GHG Efficient

Commercial aviation in the United States has a decidedly strong track record that is often overlooked or misstated. U.S. commercial aviation contributes about two percent of domestic U.S. GHG emissions.[2] To put that into context, with passenger vehicles (cars and light duty trucks) alone accounting for over 17.5 percent,[3] as illustrated in Figure 1, road transport accounts for over a fourth of U.S. GHG emissions and power plants account for over a third.[4] The picture is similar when viewed on a global basis. Worldwide commercial aviation contributes about three percent of man-made GHGs.[5] To put this into perspective, cattle and other livestock account for approximately 18 percent.[6] Further, EPA data confirm that while the overall transportation sector accounts for more than 58 percent of the nation's NO_x inventory, aviation contributes only 0.5% of that inventory.[7]

At the same time, commercial aviation is critically important to local, national and global economies, enabling a large percentage of U.S. economic output. A March 2006 study by the Campbell-Hill Aviation Group found that "the national economy is highly dependent on commercial aviation, which is directly or indirectly responsible for 5.8 percent of gross output, 5.0 percent of personal earnings and 8.8 percent of national employment."[8] The study further noted that this translated into $380 billion in earnings, 11.4 million jobs and $1.2 trillion in U.S. output in 2004. Placing our economic output side-by-side with our GHG output, it is clear that commercial aviation is an extremely GHG-efficient economic engine, bringing good "bang" for our GHG "buck."

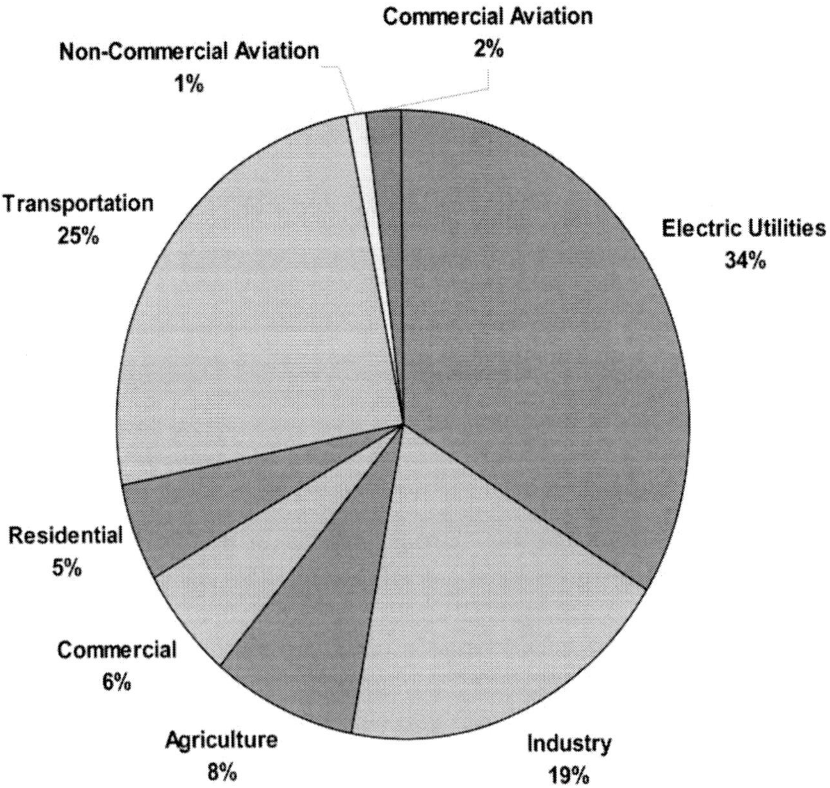

Source: U.S. EPA Data 2005

Figure 1. U.S. Aviation Greenhouse Gas Emissions 2 Percent of the Inventory

We have been able to deliver such strong economic output while reducing our emissions by continually improving our fuel efficiency through reinvestment in technology and more fuel-efficient operations. In fact, U.S. commercial airlines (passenger and cargo combined) improved their fuel efficiency by 110 percent between 1978 and 2007, which (given the one-to-one relationship between fuel consumption and carbon dioxide (CO_2)) has resulted in 2.5 billion metric tons of CO_2 savings – roughly equivalent to taking 18.7 million cars off the road each of those years.[9] Further, Bureau of Transportation Statistics data confirm that U.S. carriers burned four percent less fuel in 2006 than they did in 2000, [10] resulting in absolute reductions in GHG emissions, even though they carried 12 percent *more* passengers and 22 percent *more* cargo.

Commercial aviation's GHG efficiency compares very favorably to other modes and other sectors. While commercial aviation improved its per-passenger fuel efficiency from 1990, freight trucks showed the reverse trend, with GHG emissions growing faster than vehicle miles traveled.[11] EPA also has confirmed that passenger vehicles have lagged far behind aircraft in fuel and GHG efficiency.[12] (*See* Figure 2). Within the aviation sector, it is important to remember that different types of commercial aircraft have vastly different impacts on the environment. Commercial jets are five to six times more fuel efficient than corporate jets. The math is simple: carrying 200 people and cargo across the country in a single plane burns a lot less fuel than 33 separate corporate jets, each flying six people.

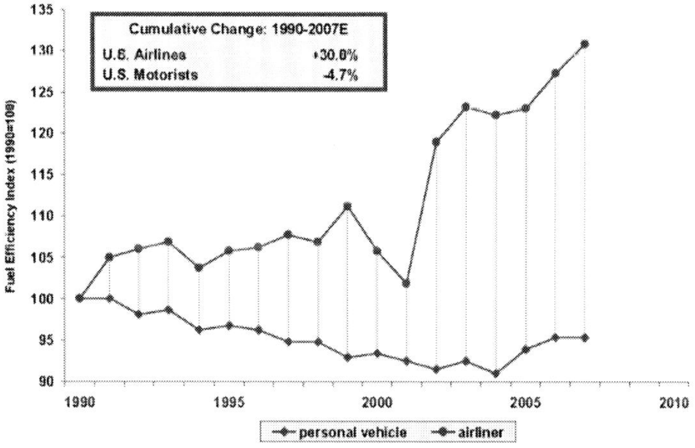

Figure 2. In Contrast to Personal Vehicles, Airline Fuel Efficiency Has Improved Substantially Since 1990

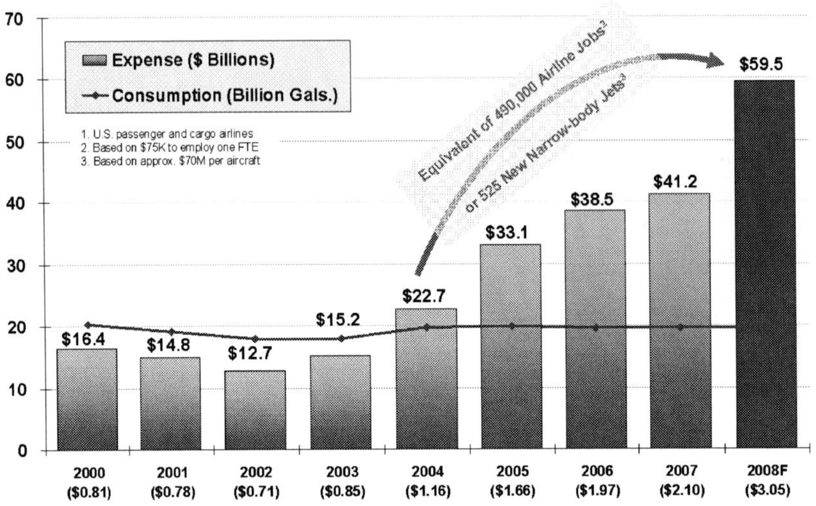

Note: Value in parentheses below year is average price paid excluding taxes, into-plane fees, pipeline tariffs and hedging costs

Sources: ATA, Energy Information Administration, Department of Transportation

Figure 3. 2008 Jet Fuel Expense Will Break 2007 Record
Avg. Price Paid per Gallon (Excl. Taxes and Into-Place Fees) Likely to Exceed $3

U.S. airlines are highly motivated to continue this trend. Fuel, long one of the two highest costs for airlines, is now our largest cost center, averaging between 30 and 50 percent of total operating expenses. In fact, jet fuel costs to the U.S. airlines in 2008 are projected to be just under $60 billion, breaking the 2007 record of $41.2 billion, resulting in what some analysts are likening to the economic effects of the 9/11 terrorist attacks.[13] As shown in Figure 3, the price change alone between 2004 and year-end 2008 is the equivalent of 490,000 airline jobs or the purchase price of 525 new narrow-body jets.

And contrary to popular belief, the airlines cannot pass on significant portions of these costs. Indeed, as illustrated in Figure 4, today's U.S. domestic air fares remain below 2000 levels, although fuel prices have tripled. While a slightly more robust international aviation market has allowed today's systemwide fares to increase approximately three percent above 2000 levels, this hardly makes up for the three-fold increase in fuel prices over the same period. (*See* Figure 5.) Thus, the market already is sending the commercial airlines a "price signal" that some call for in legislation. We have an unrelenting economic imperative to reduce fuel consumption; therefore, an economic win is an environmental win.

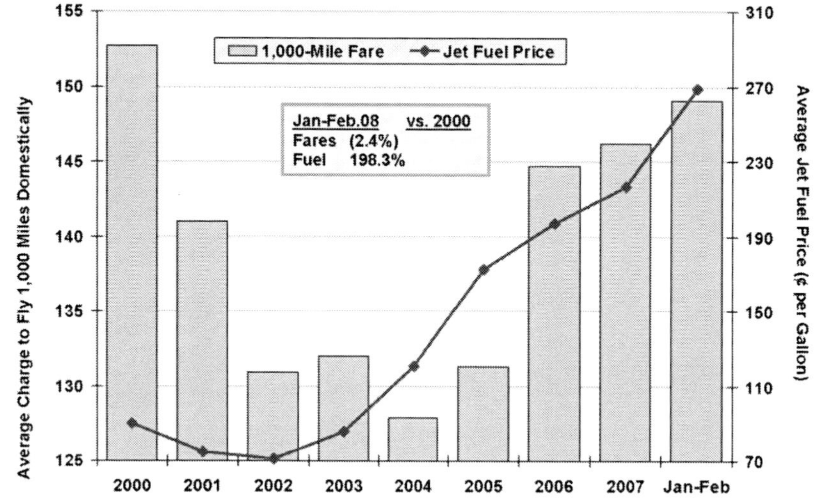

Source: ATA monthly passenger revenue report (DOM + EXP) and U.S. Energy Information Administration

Figure 4. As of Early 2008, Domestic Airfares Remain Below 2000 Levels, While Jet Fuel Prices Have Tripled

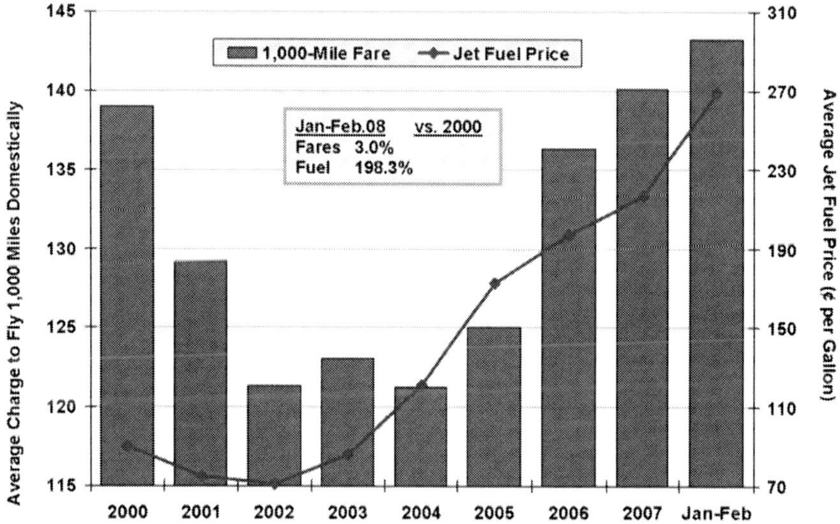

Source: ATA monthly passenger revenue report (DOM + EXP) and U.S. Energy Information Administration

Figure 5. As of Early 2008, Systemwide Airfares Just Above 2000 Levels, While Jet Fuel Prices Have Tripled

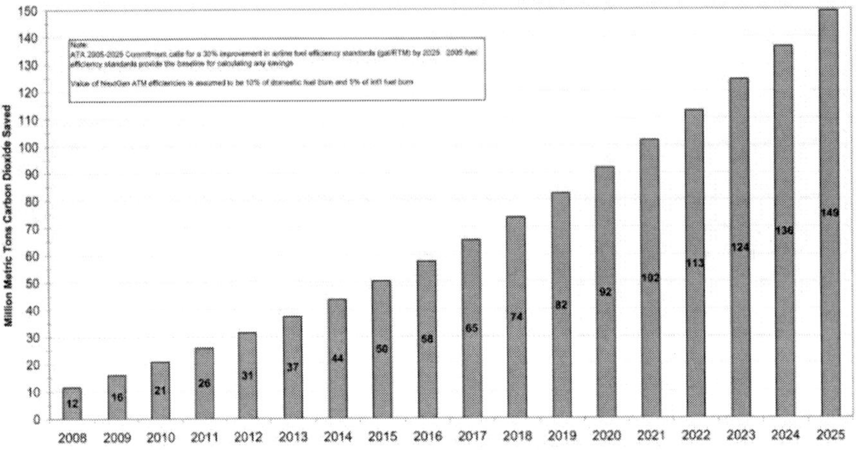

Figure 6. ATA's 30 Percent Fuel Efficiency Goal Will Translate into CO2 Savings.

ATA Airlines Are Proactively Committed to Further Limiting Their Emissions Footprint

As demand for air passenger and cargo services grows, some growth in aviation emissions is predicted. However, this growth must be kept in context. The Intergovernmental Panel on Climate Change (IPCC), which is considered the authority on this issue, has determined that under the most likely scenario, CO_2 from global aviation in 2050 will account for only about three percent of total man-made CO_2 emissions and that aviation's overall GHG impact will be around five percent.[14] Yet even though those remain relatively small numbers, the ATA carriers are relentlessly pursuing measures to further limit their emissions footprint.

At the core of our efforts, the ATA carriers have made a commitment to achieve an additional 30 percent systemwide fuel efficiency improvement through 2025, on top of prior improvements. That equates to an additional 1.2 billion metric tons of CO_2 saved – roughly equivalent to taking over 13 million cars off the road each year. (*See* Figure 6). To accomplish this, our airlines will continue and step-up the tremendous investments in new equipment and in operational innovations that have allowed us to attain such great fuel efficiency improvements in the past, achieving not only GHG emissions reductions but also reductions in local emissions such as NO_x. We are leaving no stone unturned. Some examples of our efforts include:

- **Upgrading Fleets**. Even in the highly constrained financial environment we now find ourselves in, the ATA airlines are expending billions to upgrade their fleets through investments in new airframes and engines, removing less fuel-efficient aircraft from their fleets, installing winglets to reduce drag, altering fan blades and other measures aimed at improved aerodynamics. As a critical element of our commitment to achieve an additional 30 percent fuel efficiency improvement by 2025, Boeing estimates that the North American carriers will spend approximately $730 billion on new aircraft through 2026.[15]

- **Introduction of Innovative, Cutting-Edge Technologies**. Our airlines also are investing millions of dollars in technologies to make existing airframes more efficient. For example, the airlines have undertaken equipage for Required Navigation Performance (RNP) approach procedures, which provide navigation capability to fly a more precise path into an airport. The ATA airlines also have developed software to analyze flight paths and weather conditions, allowing aircraft to fly more direct, efficient routes (subject to air traffic approval).

- **Improved In-Flight Operations**. The ATA airlines are doing all they can within the existing ATM system to utilize systems to optimize speed, flight path and altitude, which not only reduces fuel consumption and emissions in the air, but avoids wasting fuel waiting for a gate on the ground. In addition to pursuing the use of RNP approach procedures at additional locations, the ATA carriers have worked with FAA to pioneer protocols for continuous descent approaches (CDAs), which reduce both emissions and noise, and we are doggedly pursuing implementation of CDAs where the existing ATM system allows.[16] Further, our carriers are implementing Automatic Dependent Surveillance Broadcast (ADS-B) satellite tracking technology, which avoids the circuitous routings that occur with today's radar-based systems. Demonstrating that the efforts extend to the smallest details of airline operation, our members also have worked on redistribution of weight in the belly of aircraft to improve aerodynamics and have introduced life vests on certain domestic routes, allowing them to overfly water on a more direct route.

- **Improved Ground Operations**. The ATA airlines also are introducing single-engine taxiing when conditions permit, redesigning hubs and schedules to alleviate congestion and converting to electric ground support equipment when feasible. Further, they are improving ground operations by plugging into electric gate power where available to avoid running auxiliary power units and using tugs to position aircraft where feasible.

- **Reducing Onboard Weight**. The ATA airlines continue to exhaustively review ways, large and small, to reduce aircraft weight – removing seat-back phones, excess galley equipment and magazines, introducing lighter seats and beverage carts, stripping primer and paint and a myriad of other detailed measures to improve fuel efficiency.

Second, recognizing that improving fuel efficiency with today's carbon-based fuel supply can only take us so far, ATA and its airlines are making extensive resource commitments to stimulate the development of commercially viable, environmentally friendly alternative fuels. As a framework for doing this, we are a founding and principal member of the Commercial Aviation Alternative Fuels Initiative (CAAFI), a consortium of airlines, government, manufacturers, fuel suppliers, universities, airports and other stakeholders who hold the various keys to research, development and responsible implementation of alternative jet fuels. Developing alternative jet fuels is a "higher hurdle" than developing alternative fuels for ground-based units, as jet fuel must meet rigorous FAA specifications, which include reliability and stability at altitude and in greatly varying temperature and pressure conditions to ensure safety. Thus, absent a cooperative initiative like CAAFI, fuel providers almost certainly would not undertake the investments needed to clear this higher hurdle, opting instead for the surer payoff at ground level.

While each entity involved in CAAFI has a role to play, our airlines understand that – as end users of the ultimate product – they must not only make clear their specifications for alternative jet fuels, but also signal the market that we will financially back fuels meeting those specifications. On Earth Day this year, the ATA Board of Directors took another significant step in this regard, issuing the "ATA Alternative Fuels Principles Document." Among other things, that document stipulates that the ATA carriers require that any future alternative jet fuel be more environmentally friendly, on a life-cycle basis, than the jet fuel available today. Through CAAFI and other partnerships, we are undertaking the work to be sure that tomorrow's alternative jet fuel meets that criterion. And accomplishing that will ensure the full decoupling of growth in aviation demand from growth in GHG emissions.

Third, while the ATA airlines are doing all that they can to promote efficiencies within the current ATM system, the limitations of that system account for 10-15 percent of unnecessary fuel burn and resulting emissions. To address this, and to achieve much-needed modernization of our outdated ATM system, ATA and its carriers are working with FAA and other agencies on a fundamental redesign of the system through the Next Generation Air Transportation System (NextGen) project and on various regional airspace design initiatives. ATA is supporting this modernization initiative through our "Smart Skies" program.[17] However, congressional approval, including fair and equitable distribution of costs among all system users, is needed before significant progress can be made in implementing this system. Congressional authorization and implementation of this initiative will bring 10-15 percent additional savings *on top* of the ATA 30 percent commitment. (*See* Figure 7).

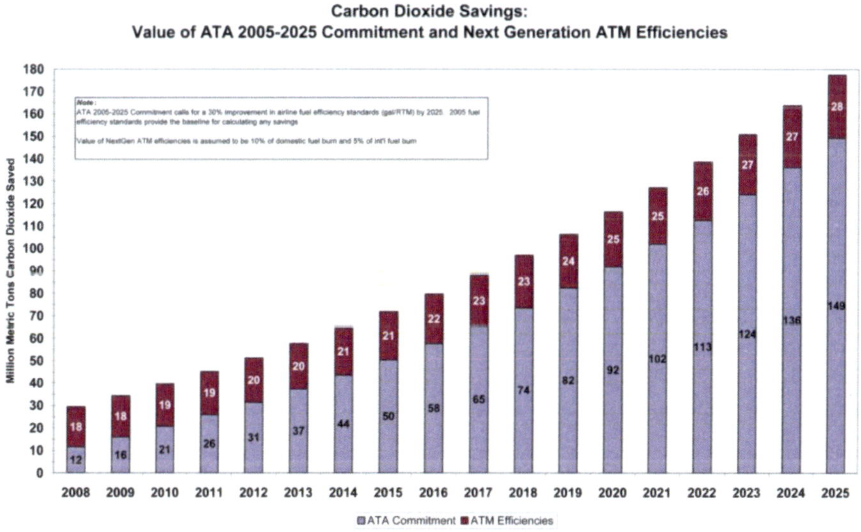

Figure 7. CO_2 Saved Under ATA and NextGen Initiatives (As if NextGen Implemented in X Year).

Fourth, at the same time ATA and its members are pushing the envelope with existing technology, we continue to contribute to work that will advance new technology. For example, ATA participates in key, joint government/stakeholder initiatives, including the Steering Committee of the Partnership for AiR Transportation Noise & Emissions Reduction (PARTNER) and the Environment and Energy Subcommittee of FAA's Research Engineering and Development Advisory Committee. While additional evolutionary environmental improvements are in the pipeline as a result of such initiatives, revolutionary environmental breakthroughs can only come about through the reinstatement of significant federal investments in basic aeronautics research and development programs at NASA and FAA. Indeed, Pratt & Whitney's new geared turbofan engine, which offers both noise and emissions benefits, as well as many features of Boeing's more environmentally efficient 787 were spawned through such programs. As we have noted in other contexts, however, congressional funding to NASA and FAA for aeronautics research and development – specifically including for environmental projects – has been cut significantly (by about 50 percent) in the past 8-10 years, compromising the public-private partnership for exploring and bringing to market products with significantly improved environmental performance.[18] Thus, we continue to urge Congress to provide this needed funding, which also is critical to preserving America's competitiveness in aeronautics.

Congress Has a Positive, Partnering Role to Play

We are confident that the measures ATA is undertaking and supporting will continue to limit and reduce aviation's emissions footprint, such that commercial aviation will remain a very small source of GHG and other emissions, even as air traffic grows with future improvements in the economy. However, as you, Mr. Chairman, recognized in your recent Op-Ed in *The Hill* publication, Congress has a key role to play. First, as noted, congressional approval for implementation of a modernized ATM system is critical, as is reinstatement of funding for research and development programs to foster aviation environmental technology breakthroughs. Further, while Congress generally is supporting several alternative fuel research programs, specific support and funding should be provided for the development of environmentally friendly alternative jet fuels.

Just as we ask Congress to continue to work with us, we also urge Congress to calibrate any climate change-related legislation so it does not work against our efforts. To have the resources to continue our fuel efficiency and other advances, we must have the capital to invest in newer aircraft and other emissions-reducing measures. Indeed, FAA estimates that 90 percent of the fuel efficiency and emissions improvements that the airlines have achieved has come through the airlines' own investments in technology. Punitive economic measures that siphon funds out of our industry would severely threaten our ability to continue that record.

In constant year dollars, NASA funding for aeronautics research was cut by about one-third between 1998 and 2000, reducing the breadth of ongoing research and prompting NASA to establish research programs with reduced goals, particularly with regard to TRL (technology readiness level). This significantly reduces the likelihood that the results of NASA research will find their way into the marketplace in a timely manner, if at all. The ultimate consequence is that the federal expenditures are inconsistent with the longterm goal of support for an aviation enterprise compatible with national goals for environmental stewardship.

Against this backdrop, we are compelled to share our concerns about the apparent front-runner cap-andtrade legislation in the U.S. Senate, S. 2191, the "Lieberman-Warner Climate Security Act," in the hopes that the House of Representatives will craft its legislation to avoid or minimize those concerns.

First, the Lieberman-Warner bill would, in effect, impose a punitive emissions tax on aviation, which would not only harm the economy but also would be counterproductive. As drafted, the bill proposes to cover the transportation sector – including aviation – indirectly, through a cap-and-trade system "upstream," which would require fuel producers to acquire allowances sufficient to cover the GHG content of the fuel they sell to the transport sector. Fuel producers will incorporate the cost of these allowances into fuel prices, passing the costs on to fuel consumers (including airlines) – in effect, operating as a fuel tax on jet fuel and other transportation fuels. This would have significant economic repercussions on the airline industry and the economy, as every penny increase in the price of a gallon of jet fuel drives an additional $ 190-200 million in annual fuel costs for U.S. airlines.

It is not difficult to calculate the likely costs of application of the Lieberman-Warner bill to aviation. Unlike most sectors, commercial aviation is required to report all of its fuel consumption to the federal government, which compiles and reports this data. Based on this data, and factoring in FAA forecast information, the annual costs to the U.S. commercial airlines of the

Lieberman-Warner bill in 2012 would be approximately $5 billion, assuming a $25 emissions allowance price. Using analysts' estimates that emissions allowance prices likely will be in the $40 range by 2020, the annual costs to aviation would escalate to almost $10 billion in that year, and would grow thereafter. (*See* Figure 8). These increased costs would diminish the airlines' ability to continue to realize the tremendous fuel efficiency improvements and emissions reductions we have achieved within the industry and, therefore, would be counterproductive. Indeed, it is difficult to imagine how we could handle a GHG-based surcharge on top of the exorbitant fuel prices we are experiencing.

Second, based on our fuel and GHG efficiency records and commitments, application of a cap-and-trade bill to commercial aviation simply is unnecessary. As noted, we already are incentivized by the market to minimize GHGs, without further market-based measures. However, if such a measure is to be applied to aviation, it should be carefully calibrated to take key considerations into account, which the Lieberman- Warner bill currently does not do.

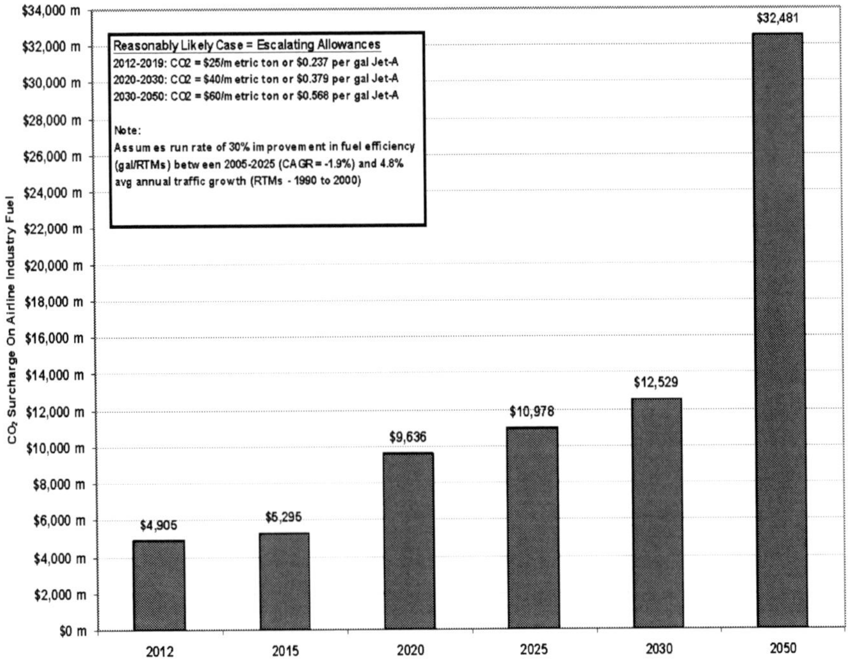

Figure 8. Lieberman–Warner Bill – Fuel Surcharge Costs to U.S. Commercial Airlines.

One such mechanism would be to provide the commercial airlines with allowances up front, either directly or as a *required* pass-through from fuel providers, in recognition of the fuel efficiency achievements we have made to date and the importance of preserving the airlines' ability to continue to invest in new aircraft technology. As drafted, the Lieberman-Warner bill does not do this. In contrast, the bill would accord to several sectors – including to industries that do not come anywhere near commercial airlines' fuel and GHG efficiency record – a tremendous amount of free allowances, purportedly to cushion the economic blow and to allow them to invest in modernizing their equipment and facilities to reduce emissions. In effect, the bill would require the airlines to subsidize future efforts of other industries that have done comparatively little to reduce their GHG profiles. The U.S. House of Representatives can avoid the inequity and public policy flaws in this approach in crafting its own legislation.

Another key calibration mechanism would be to take some of the proceeds generated from the auctioning of allowances and reinvest those proceeds into aviation. This could allow for additional funding of programs and technologies that promise to further reduce aviation GHG emissions. With a 10-15 percent GHG savings directly on the line, equipage for NextGen is perhaps the most significant candidate in this regard, but funding for aviation alternative fuel and aircraft environmental technology breakthroughs are also well-deserving candidates. A fundamental flaw of the Lieberman-Warner bill is that while it proposes to rechannel proceeds from auctions into industries like automobile manufacturing, it does not include any provisions for reinvestment in aviation.

Further, any climate change legislation proposing to cover aviation should be crafted to take into account the international nature of aviation, not only that aviation is a global industry and that U.S. carriers must compete with the airlines of other nations on many routes, but also that the United States by treaty has agreed that the International Civil Aviation Organization (ICAO) has the authority to establish standards and policy for international flights.[19] Arguably, the United States should defer to ICAO for additional measures addressing aviation GHGs. At a minimum, however, we should ensure that any measures taken in the U.S. are compatible with our international aviation agreements.

As an additional example of the need to carefully calibrate any climate change legislation, it is important to recognize that policies that make flying more expensive can have the effect of pushing more people into their cars. This would result not only in increased GHG emissions from the less fuel-efficient ground transportation sector, but also in more GHG emissions and increased traffic deaths, as the highways are much less safe than the air. Again, the U.S. House of Representatives has the opportunity to factor such concerns into its work on this issue.

CONCLUSION

I close by asking you to note the achievements that commercial airlines have made in reducing fuel burn and GHGs, particularly when compared to other industries, and the actions that we are taking to continue our progress in this regard. While we are fully committed to working with Congress and are asking for congressional leadership and support in each of the areas I have described, we are not asking you to work for us, we're asking you to work with us in addressing this environmental and energy concern. We also are urging you to refrain from adopting policies that would work against our efforts. A vibrant, competitive and growing aviation sector is a key part of the solution, not an impediment to ensuring a future where a strong economy, freedom from foreign oil and cleaner air are the order of the day.

End Notes

[1]ATA airline members include ABX Air, AirTran Airways, Alaska Airlines, Aloha Airlines, American Airlines, AS TAR Air Cargo, Atlas Air, Continental Airlines, Delta Air Lines, Evergreen International Airlines, Federal Express Corporation, Hawaiian Airlines, JetBlue Airways, Midwest Airlines, Northwest Airlines, Southwest

[2] The United States Environmental Protection Agency's (EPA's) most recent general inventory reports commercial aviation's contribution to the total GHG emissions in 2006 was 2.04 percent. EPA, *Inventory of Greenhouse Gas Emissions and Sinks: 1990 -2006* (April 15, 2008) (hereinafter *EPA GHG Inventory 1990-2006)* at pages ES-4 and 21 ("in 2006, total U.S. greenhouse gas emissions were 7,054.2" teragrams of carbon dioxide equivalent (Tg CO2 Eq)) and Table 2-15 at pp. 2-22 & 2-23 ("Commercial Aircraft – Domestic" account for 143.6 Tg. CO2 Eq.).

[3] *EPA GHG Inventory 1990-2006*, Table 2-15 at pp. 2-22 & 2-23.

[4] *EPA GHG Inventory 1990-2006.*

[5] It is estimated that on a worldwide basis, commercial aviation accounts for approximately three percent of total GHGs, while at the same time contributing over eight percent of the world's economic activity. *See* International Air Transport Association, *Debunking Some Persistent Myths about Air Transport and the Environment.*

[6] United Nations, Livestock Environment and Development Initiative, *Livestock's Long Shadow – Environmental Issues and Options* (2006) at p. 271.

[7] EPA data cited in *Overview of Aviation & Air Quality Impacts* (presentation by Dr. Lourdes Maurice, FAA, March 19, 2008).

[8] The Campbell-Hill Aviation Group, *Commercial Aviation and the American Economy*, March 2006.

[9] Between 1978 and 2006, our carriers' fuel efficiency improvement was 103 percent. Our 2007 data shows a jump to 110 percent.

[10] This figure reflects reductions achieved by all international and domestic flights from U.S. carriers. The reductions in domestic flights were even more dramatic, as the U.S. carriers burned 13.4 percent less fuel on domestic flights since 2000. EPA GHG Inventory 1990-2006, Table A-98 at pp. A-123 & A-124.

[11] *EPA GHG Inventory 1990-2006* at 3-8.

[12] *Id.*

[13] *See* J.P. Morgan Securities North America Corporate Research (April 15, 2008).

[14] IPCC, *Aviation and the Global Atmosphere* (1999) at 8.

[15] The Boeing Company (2008).

[16] For example, one of our carriers is achieving an average savings of 1300 pounds of CO_2 savings per flight for approaches into the Atlanta airport.

[17] "Smart Skies" is a national campaign led by ATA airlines, which advocates modernization of the U.S. ATM system and its funding mechanisms. For more on this initiative, see the Smart Skies Web site, at http://www.smartskies.org.

[18] While later funding cuts were even more drastic, a 2002 study by the National Academy of Sciences observed:

In constant year dollars, NASA funding for aeronautics research was cut by about one-third between 1998 and 2000, reducing the breadth of ongoing research and prompting NASA to establish research programs with reduced goals, particularly with regard to TRL (technology readiness level). This significantly reduces the likelihood that the results of NASA research will find their way into the marketplace in a timely manner, if at all. The ultimate consequence is that the federal expenditures are inconsistent with the long-term goal of support for an aviation enterprise compatible with national goals for environmental stewardship.

See National Academy of Sciences, Committee on Aeronautics Research and Technology for Environmental Compatibility, *For Greener Skies: Reducing Environmental Impacts of Aviation* at 44 (2002).

[19] This is pursuant to the Convention on International Civil Aviation, commonly referred to as the "Chicago Convention," to which 190 countries, including the United States, are parties.

In: Aviation and Climate Change
Editors: George T. Blumenthal

ISBN: 978-1-60876-757-1

Chapter 8

AVIATION AND THE ENVIRONMENT: NEXTGEN AND RESEARCH AND DEVELOPMENT ARE KEYS TO REDUCING EMISSIONS AND THEIR IMPACT ON HEALTH AND CLIMATE

Gerald L. Dillingham

WHY GAO DID THIS STUDY

Collaboration between the federal government and the aviation industry has led to reductions in aviation emissions, but growing air traffic has partially offset these reductions. The Federal Aviation Administration (FAA), together with the National Aeronautics and Space Administration (NASA), the Environmental Protection Agency (EPA), and others, is working to increase the efficiency, safety, and capacity of the national airspace system and at the same time reduce aviation emissions, in part, by transforming the current air traffic control system to the Next Generation Air Transportation System (NextGen). This effort involves new technologies and air traffic procedures that can reduce aviation emissions and incorporates research and development (R&D) on emissions-reduction technologies. Reducing aviation emissions is important both to minimize their adverse health and environmental effects and to alleviate public concerns about them that could constrain the expansion of airport infrastructure and aviation operations needed to meet demand.

This testimony addresses (1) the scope and nature of aviation emissions, (2) the status of selected key federal efforts to reduce aviation emissions, and (3) next steps and challenges in reducing aviation emissions. The testimony updates prior GAO work with FAA data, literature reviews, and interviews with agency officials, industry and environmental stakeholders, and selected experts.

What GAO Found

Aviation contributes a modest but growing proportion of total U.S. emissions, and these emissions contribute to adverse health and environmental effects. Aircraft and airport operations, including those of service and passenger vehicles, emit ozone and other substances that contribute to local air pollution, as well as carbon dioxide and other greenhouse gases that contribute to climate change. EPA estimates that aviation emissions account for less than 1 percent of local air pollution nationwide and about 2.7 percent of U.S. greenhouse gas emissions, but these emissions are expected to grow as air traffic increases.

Two key federal efforts, if implemented effectively, can help to reduce aviation emissions—NextGen initiatives in the near term and research and development over the longer term. For example, NextGen technologies and procedures, such as satellite-based navigation systems, should allow for more direct routing, which could improve fuel efficiency and reduce carbon dioxide emissions. Federal research and development efforts—led by FAA and NASA in collaboration with industry and academia—have achieved significant reductions in aircraft emissions through improved aircraft and engine technologies, and federal officials and aviation experts agree that such efforts are the most effective means of achieving further reductions in the longer term. Federal R&D on aviation emissions also focuses on improving the scientific understanding of aviation emissions and developing lower-emitting aviation fuels.

Source: FAA.

Sources of Aviation Emissions

Next steps in reducing aviation emissions include managing NextGen initiatives efficiently; deploying NextGen technologies and procedures as soon as practicable to realize their benefits, including lower emissions levels; and managing a decline in R&D funding, in part, by setting priorities for R&D on NextGen and emissions-reduction technologies. Challenges in reducing aviation emissions include designing aircraft that can simultaneously reduce noise and emissions of air pollutants and greenhouse gases; encouraging financially stressed airlines to purchase more fuel-efficient aircraft and emissions-reduction technologies; addressing the impact on airport expansion of more stringent EPA air quality standards and growing public concerns about the effects of aviation emissions; and responding to proposed domestic and international measures for reducing greenhouse gases that could affect the financial solvency and competitiveness of U.S. airlines.

Mr. Chairman and Members of the Subcommittee:

I appreciate the opportunity to testify before you on aviation emissions, one of the key sources of concern about the environmental effects of aviation. Over the past 30 years, the federal government, the aviation industry, and other private parties have worked collaboratively to achieve steady reductions in aircraft emissions.[1] Nevertheless, increases in air traffic, which have enhanced the nation's productivity and mobility, have partially offset these reductions, as more flights have produced more emissions and congestion has led to flight delays. According to the Federal Aviation Administration (FAA), this growth in air traffic will continue, with the number of flights increasing 20 percent by 2015 and 60 percent by 2030.[2] In light of these developments, concerns about the environmental effects of aviation emissions have persisted. Moreover, better scientific understanding of the potential health effects of certain aviation emissions and their contribution to climate change has intensified the public's concerns.

To accommodate the expected growth in air traffic, FAA is leading a multipronged, multiagency effort to increase the efficiency, safety, and capacity of the national airspace system. This effort includes transforming the current air traffic control system into the Next Generation Air Transportation System (NextGen)[3] and will require airport and runway expansion. The NextGen initiative incorporates research and development (R&D) on emissions-reduction technologies, alternative fuels, and cleaner and quieter air traffic management procedures. This R&D is necessary both to meet anticipated domestic and international environmental standards and to reduce the environmental impact of aviation. Meeting environmental standards can limit the adverse effects of aviation emissions on air quality and climate, and addressing public concerns about aviation emissions is necessary to avoid constraints on the expansion of aviation operations and airport infrastructure planned under NextGen.[4]

Under the National Environmental Policy Act of 1969, agencies evaluate the likely environmental effects of projects they are proposing using an environmental assessment or, if the projects likely would significantly affect the environment, a more detailed environmental impact statement.[5] FAA typically carries out one of these evaluations for federally financed airport construction projects, including the construction of federally subsidized runways. In addition, under the Clean Air Act's conformity provision, no federal agency may approve or provide financial assistance for any activity that does not conform to an applicable state implementation plan.[6] Therefore, FAA must evaluate whether a proposed federal action associated with an airport project conforms with the applicable state implementation plan before approving or funding the project.[7] In addition, the Clean Air Act mandates standards for mobile sources of emission, such as aircraft and the equipment that service them at airports. EPA sets emissions standards for aircraft and has chosen to adopt international emissions standards for aircraft set by the International Civil Aviation Organization (ICAO).[8]

As requested, my testimony today focuses on aviation emissions. It will address the following questions: (1) What are the scope and nature of aviation emissions? (2) What is the status of selected key federal efforts to address aviation emissions? and (3) What are some next steps and major challenges for the federal government, the aviation industry, and Congress related to aviation emissions? My statement is based on previous GAO reports[9] updated with a synthesis of recent empirical literature and interviews with officials from FAA, the National Aeronautics and Space Administration (NASA), and the U.S. Environmental Protection Agency (EPA); representatives of aviation industry and environmental associations, and selected aviation emissions experts.[10] We balanced the selection of these experts to capture the views of the many different groups involved in aviation emissions reduction efforts and NextGen. We conducted our work from March to May 2008 in accordance with generally accepted government auditing standards. Those standards require that we plan and perform the study to obtain sufficient, appropriate evidence to provide a reasonable basis for our findings and conclusions based on our study objectives. We believe that the evidence obtained provides a reasonable basis for our findings and conclusions based on our study objectives.

SUMMARY

Currently, aviation contributes a modest proportion of total emissions in the United States, but its share could increase in the future, and aviation emissions can have a detrimental effect on health and the environment. Aircraft are the primary source of aviation emissions, but airport operations, including those of service and passenger vehicles, also produce emissions. Together, aircraft operations in the vicinity of the airport and other airport sources emit nitrogen oxides, which lead to the formation of ground-level ozone (also known as smog), and other substances that contribute to local air pollution, as well as carbon dioxide and other greenhouse gases that rise into the atmosphere and contribute to climate change. Aircraft operations in the upper atmosphere are, however, the primary aviation-related source of greenhouse gas emissions. Currently, according to EPA estimates, aviation emissions account for less than 1 percent of local air pollution nationwide and about 2.7 percent of U.S. greenhouse gas emissions. This proportion is, however, expected to grow with projected increases in air traffic, despite expected improvements in fuel efficiency. Notably, according to FAA, emissions of nitrogen oxides from aviation sources will increase by over 90 percent by 2025 if not addressed. This increase is likely to increase ozone, which aggravates respiratory ailments. Increases in air traffic also mean increases in carbon dioxide emissions and increases in aviation's contribution to climate change, according to the International Panel on Climate Change (IPCC).

Two key federal efforts, if implemented effectively, can help to reduce aviation emissions—near-term NextGen initiatives and R&D over the longer term to fully enable NextGen and reduce aircraft emissions. Some NextGen technologies and procedures, such as satellite-based navigation systems, should allow for more direct routing, which could improve fuel efficiency and reduce carbon dioxide emissions. According to FAA, the full implementation of NextGen could reduce greenhouse gas emissions from aircraft by up to 12 percent by 2025. Federal R&D efforts—led primarily by FAA and NASA and often conducted in collaboration with industry and academia—have achieved significant reductions in aircraft emissions over the last 30 years, and FAA and NASA officials and aviation experts agree that such efforts are the most effective means of achieving further reductions in the longer term. As part of the a national plan for aeronautics R&D, issued by the White House Office of Science and Technology Policy, the federal government supports a comprehensive approach to R&D on aviation emissions involving FAA, NASA, and other federal agencies that is intended both to improve scientific understanding of the impact of aviation emissions and to develop new technologies, fuels, and air traffic management approaches. Better understanding of the nature and impact of aviation emissions can inform the development of lower- emitting alternative fuels, more efficient air traffic management technologies and procedures, and more fuel-efficient aircraft engines.

Reducing aviation emissions includes steps that FAA and others can take to move the implementation of NextGen forward and to support R&D on NextGen and emissions-reduction technologies, as well as technical, financial, and regulatory challenges facing the federal government, the aviation industry, and Congress. One step for FAA is to ensure the efficiency of NextGen's management by, for example, addressing congressional leaders' and stakeholders' concerns about the program's management structure and authority. Another step for FAA is to further deploy, as soon as practicable, NextGen technologies and procedures, such as the more efficient takeoff and landing procedures now in use at a few airports, to realize their benefits and lower emissions levels. A third step, for FAA and NASA, is managing a decline in federal funding for aeronautics research, the research category that includes work on aviation emissions, new aircraft and engine technologies, and alternative fuels. As a result of this decline, NASA is now sometimes developing technologies to a lower maturity level than in the past, and the technologies are less ready for manufacturers to adopt them. The administration's reauthorization bill for FAA seeks some additional funding for an initiative that could lead to the earlier maturation of certain emissions-reduction technologies, but according to some experts, increased funding of the initiative could increase the probability of success and decrease the time needed to achieve that success. Challenges in reducing aviation emissions for the federal government, the aviation industry, and Congress include designing aircraft that can simultaneously reduce noise and emissions of air pollutants and greenhouse gases; encouraging financially stressed airlines to purchase more fuel-efficient aircraft and emissions-reduction technologies; addressing the impact on airport expansion of more stringent EPA air quality standards and growing public concerns about effects of aviation emissions; and responding to proposed domestic and international measures for reducing greenhouse gases that could affect the financial solvency and competitiveness of U.S. airlines.

AVIATION'S SMALL BUT GROWING PROPORTION OF TOTAL EMISSIONS CONTRIBUTES TO HEALTH AND ENVIRONMENTAL EFFECTS

Aviation-related activities contribute to local air pollution and produce greenhouse gases that cause climate change. Aircraft account for about 70 to 80 percent of aviation emissions, producing emissions that mainly affect air quality below 3,000 feet and increase greenhouse gases at higher altitudes. At ground level, airport operations, including those of motor vehicles[11] traveling to and from the airport, ground service equipment,[12] and stationary sources such as incinerators and boilers, also produce emissions. Together, aircraft operations in the vicinity of the airport and other airport sources produce emissions such as carbon monoxide, sulfur oxides, particulate matter, nitrogen oxides, unburned hydrocarbons, hazardous air pollutants,[13] and ozone[14] that contribute to air pollution. In addition, these sources emit carbon dioxide and other greenhouse gases that contribute to climate change, but aircraft operations in the upper atmosphere are the primary source of aviation-related greenhouse gases. Carbon dioxide is both the primary aircraft emission and the primary contributor to climate change. It survives in the atmosphere for over 100 years. Furthermore, other gases and particles emitted by aircraft—including water vapor, nitrogen oxides, soot, contrails,[15] and sulfate—can also have an impact on climate, but the magnitude of this impact is unknown, according to FAA. Figure 1 illustrates aviation's impact on air quality and climate.

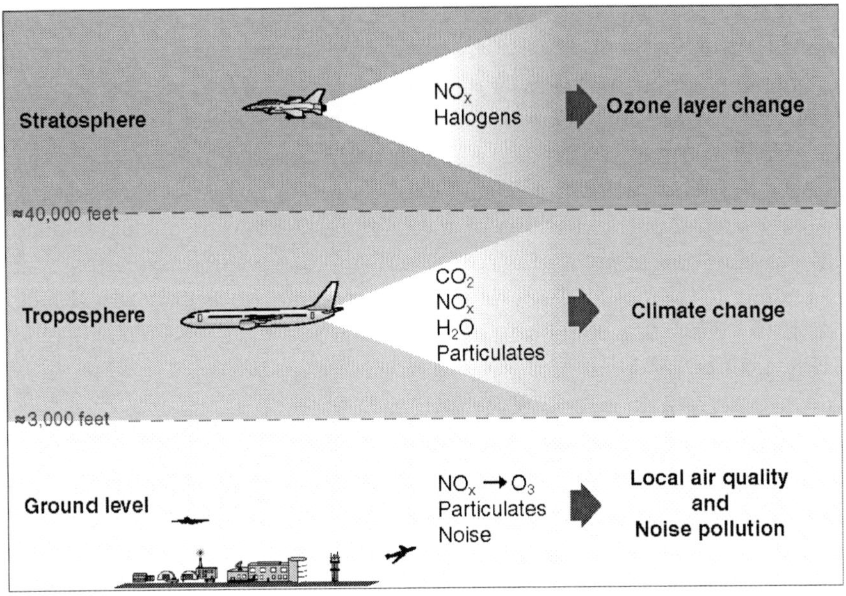

Source: GAO.

Figure 1. Environmental Effects of Aviation Emissions and Noise.

Currently, aviation accounts for a small portion of air pollutants and greenhouse gas emissions. Specifically, aviation emissions represent less than 1 percent of air pollution nationwide, but their impact on air quality could be higher in the vicinity of airports. In addition, aviation accounts for about 2.7 percent of the total U.S. contribution of greenhouse gas emissions, according to the Department of Transportation's Center for Climate Change and Environment. A 1999 study by the United Nations' Intergovernmental Panel on Climate Change (IPCC) estimated that global aircraft emissions generally accounted for approximately 3.5 percent of the warming generated by human activity.[16]

As air traffic increases, aviation's contribution to air pollution and climate change could also grow, despite ongoing improvements in fuel efficiency, particularly if other sectors achieve significant reductions. In addition, aviation's impact on air quality is changing as more fuel-efficient, quieter aircraft engines are placed in service. While new aircraft engine technologies have reduced fuel consumption, noise, and emissions of most pollutants, they have not achieved the same level of reductions in nitrogen oxide emissions, which contribute to ozone formation. According to FAA, nitrogen oxide emissions from aviation will increase by over 90 percent by 2025 without improvements in aircraft emissions technologies and air traffic management, and emissions of other air pollutants will also increase, as shown in figure 2. Additionally, aviation's greenhouse gas emissions and potential contribution to climate change is expected to increase. IPCC has estimated that aircraft emissions are likely to grow by 3 percent per year, outpacing the emissions reductions achieved through technological improvements. Furthermore, as emissions from other sources decline, aviation's contribution to climate change may become proportionally larger, according to FAA. Alternative fuels are not yet available in sufficient quantities for jet aircraft, as they are for some other uses, and therefore aviation cannot yet adopt this approach to reduce its greenhouse gas emissions (see discussion below on U.S. efforts to develop alternative fuels for aviation).

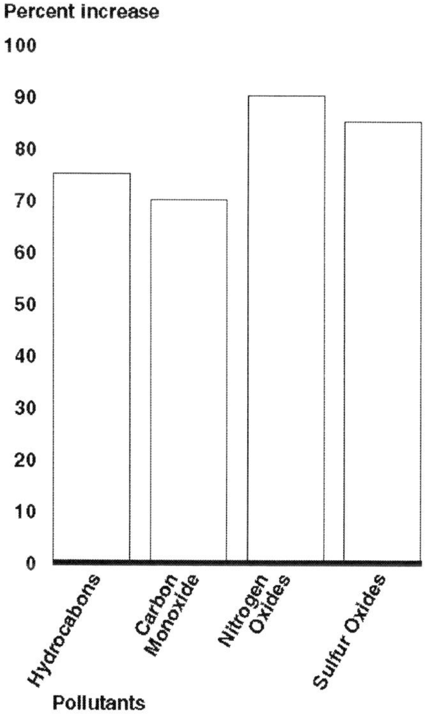

Source: FAA.

Note: According to FAA, the increases in aviation-related pollutants are baseline forecasts that do not account for potential improvements in aircraft technology and air traffic management.

Figure 2. FAA Analysis of Growth in Aviation Related Pollutants by 2025

Aviation emissions, like other combustible emissions, include pollutants that affect health. While it is difficult to determine the health effects of pollution from any one source, the nitrogen oxides produced by aircraft engines contribute to the formation of ozone, the air pollutant of most concern in the United States and other industrialized countries. Ozone has been shown to aggravate respiratory ailments. A National Research Council panel recently concluded that there is strong evidence that even short-term exposure to ozone is likely to contribute to premature deaths of people with asthma, heart disease, and other preexisting conditions. With improvements in aircraft fuel efficiency and the expected resulting increases in nitrogen oxide emissions, aviation's contribution to ozone formation may increase. In addition, aviation is associated with other air pollutants, such as hazardous air pollutants, including benzene and formaldehyde, and particulate matter, all of which can adversely affect health. Data on emissions of hazardous air pollutants in the vicinity of airports are limited, but EPA estimates that aviation's production of these pollutants is small relative to other sources, such as on-road vehicles. Nevertheless, according to EPA, there is growing public concern about the health effects of the hazardous air pollutants and particulate matter associated with aviation emissions. See appendix I for more detailed information on the health and environmental effects of aviation emissions.

Carbon dioxide and other greenhouse gas emissions from aircraft operations in the atmosphere, together with ground-level aviation emissions that gradually rise into the atmosphere, contribute to global warming and climate change. IPCC's most recent report[17] documents mounting evidence of global warming and projects the potential catastrophic effects of climate change. As figure 6 shows, climate change affects precipitation, sea levels, and winds as well as temperature, and these changes in turn will increasingly affect economies and infrastructure around the world.

KEY FEDERAL EFFORTS TO ADDRESS AVIATION EMISSIONS INCLUDE NEAR-TERM OPERATIONAL CHANGES AND LONGER-TERM R&D INITIATIVES

Two key federal efforts, if implemented effectively, can help to reduce aviation emissions—near-term NextGen initiatives and an array of R&D programs over the longer term to fully enable NextGen and to reduce aircraft emissions. The NextGen initiatives are primarily intended to improve the efficiency of the aviation system so that it can handle expected increases in air traffic, but these initiatives can also help reduce aviation emissions. In addition, the federal government, led by FAA and NASA, has longer-term R&D programs in place to improve the scientific understanding of the impact of aviation emissions in order to inform decisions about emissions-reduction strategies, explore potential emissions-reducing alternative fuels, and develop NextGen and aircraft emissions-reduction technologies.

NextGen Initiatives Have the Potential to Help Reduce Emissions

Technologies and procedures that are being developed as part of NextGen to improve the efficiency of flight operations can also reduce aircraft emissions. According to FAA, the implementation of NextGen could reduce greenhouse gas emissions from aircraft by up to 12 percent. One NextGen technology, considered a centerpiece of NextGen, is the Automatic Dependent Surveillance-Broadcast (ADS-B) satellite aircraft navigation system. ADS-B is designed, along with other navigation technologies, to enable more precise control of aircraft during en route flight, approach, and descent. ADS-B will allow for closer and safer separations between aircraft and more direct routing, which will improve fuel efficiency and reduce carbon dioxide emissions. This improved control will also facilitate the use of air traffic control procedures that will reduce communities' exposure to aviation emissions and noise. One such procedure, Continuous Descent Arrivals (CDA), allows aircraft to remain at cruise altitudes longer as they approach destination airports, use lower power levels, and thereby lower emissions and noise during landings. Figure 3 shows how CDA compares with the current step-down approach to landing, in which aircraft make alternate short descents and forward thrusts, which produce more emissions and noise than continuous descents. A limited number of airports have already incorporated CDA into their operations. For example, according to officials from Los Angeles International Airport, nearly 25 percent of landings at their airport use CDA procedures in one of the airport's standard terminal approaches. In addition, United Parcel Service plans to begin using a nighttime CDA procedure, designed and tested at the Louisville International Airport, for its hub operations.

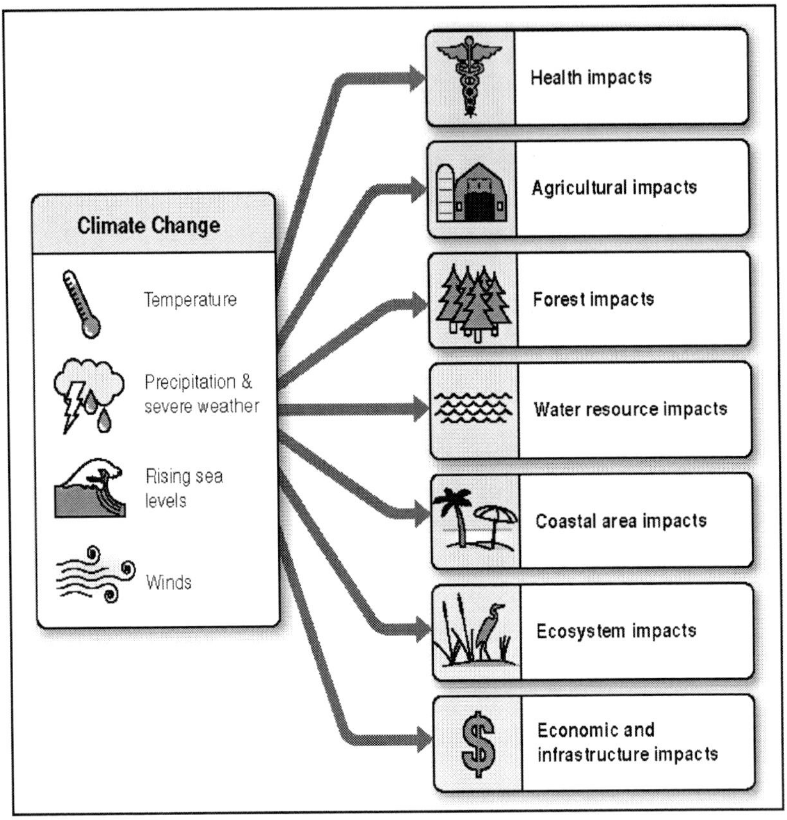

Source: EPA and FAA.

Figure 3. Concerns about the Effects of Climate Change

Two closely associated NextGen initiatives, Area Navigation (RNAV) and Required Navigation Performance (RNP), have the potential to modify the environmental impact of aviation by providing enhanced navigational capability to the pilot. RNAV equipment can compute an airplane's position, actual track, and ground speed, and then provide meaningful information on the route of flight selected by the pilot. RNP will permit the airplane to descend on a precise route that will allow it to avoid populated areas, reduce its consumption of fuel, and lower its emissions of carbon dioxide and nitrogen oxides.[18] See figure 4. Currently, over 350 RNAV/RNP procedures are available at 54 airports, including Dallas/Fort Worth, Miami International, Washington Dulles, and Atlanta Hartsfield.

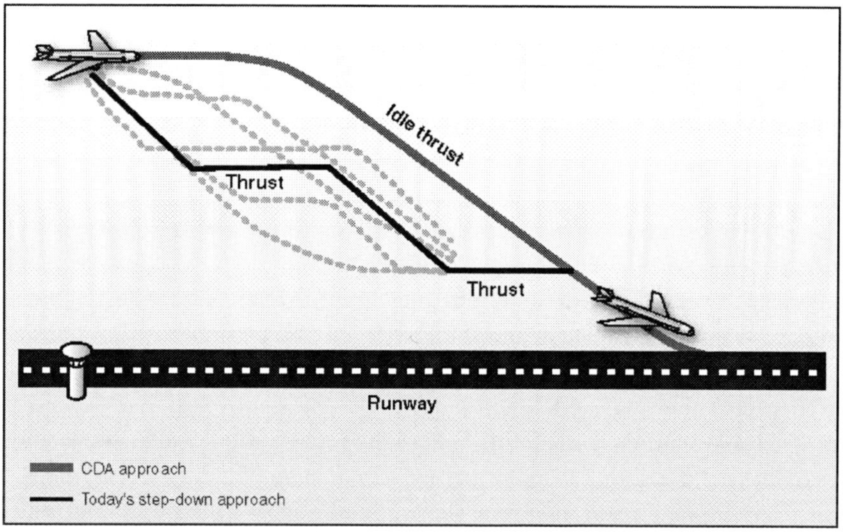

Sources: Naverus and AVTECH.

Note: Continuous Descent Arrivals keep aircraft higher for longer and have them descend at near-idle power to touchdown. Optimal profiles are not always possible, especially at busy airports.

Figure 4. Comparison of CDA and Current Step-Down Approach

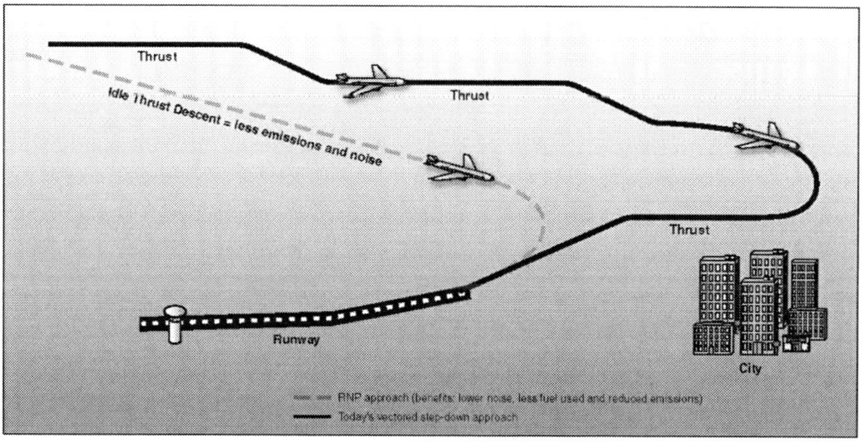

Sources: Naverus and AVTECH.

Note: An RNP approach and path allows for idle-thrust, continuous descent instead of today's step- down approaches with vectors. RNP precision and curved-approach flexibility can shift flight paths to avoid populated areas.

Figure 5. Comparison of RNP and Current Step-Down Approach.

Still another NextGen initiative, High-Density Terminal and Airport Operations, is intended to improve the efficiency of aircraft operations at busy airports, and, in the process, reduce emissions. At high-density airports, the demand for access to runways is high, and arrivals and departures take place on multiple runways. The combination of arrivals, departures, and taxiing operations may result in congestion, which in turn produces delays, emissions, and noise as aircraft wait to take off and land. Under the High-Density Terminal and Airport Operations initiative, which FAA has just begun to implement, aircraft arriving and departing from different directions would be assigned to multiple runways and safely merged into continuous flows despite bad weather and low visibility. To guarantee safe separation, these airports would need enhanced navigation capabilities and controllers with access to increased automation. Under this initiative, aircraft would also move more efficiently on the ground, using procedures that are under development to reduce spacing and separation requirements and improve the flow of air traffic into and out of busy metropolitan airspace. More efficient aircraft movement would increase fuel efficiency and reduce emissions and noise. Although the implementation of this initiative is in the early stages, FAA has identified the R&D needed to move it forward.

Technologies and procedures planned for NextGen should also help improve the efficiency of flights between the United States and other nations, further reducing emissions, particularly of greenhouse gases. A test program scheduled to begin in the fall of 2008, known as the Atlantic Interoperability Initiative to Reduce Emissions (AIRE), sponsored by FAA and the European Commission, Boeing, and Airbus, will involve gate-togate testing of improved procedures on the airport surface, during departures and arrivals, and while cruising over the ocean. Some of the procedures to be tested will use technologies such as ADS-B. A similar effort—the Asia and South Pacific Initiative to Reduce Emissions (ASPIRE)—was launched earlier this year, involving the United States, Australia, and New Zealand.

Federal R&D Focuses on Long-Term Approaches to Addressing Aviation Emissions

We have previously reported[19] that the federal government and industry have achieved significant reductions in some aircraft emissions, such as carbon dioxide, through past R&D efforts, and federal officials and aviation experts agree that such efforts are the most effective means of achieving further reductions in the longer term[20]. As part of the a national plan for aeronautics R&D, issued by the White House Office of Science and Technology Policy, the federal government supports a comprehensive approach to R&D on aviation emissions that involves FAA, NASA, and other federal agencies. According to FAA, this approach includes efforts to improve the scientific understanding of the nature and impact of aviation emissions and thereby inform the development of more fuel-efficient aircraft, of alternative fuels that can reduce aircraft emissions, and of air traffic management technologies that further improve the efficiency of aviation operations. NASA, industry, and academia are important partners in these efforts. Notably, however, the development of breakthrough technologies, such as highly fuel-efficient aircraft engines that emit fewer greenhouse gases and air pollutants, is expensive and can take a long time, both to conduct the research and to implement the new technologies in new aircraft designs and introduce these new aircraft into the fleet. Successfully developing these technologies also requires the support and cooperation of stakeholders throughout the aviation industry.

FAA Supports Research on Improving the Scientific Understanding of Aviation Emissions and on Alternative Fuels

Improving the scientific understanding of aviation emissions can help guide the development of approaches to reducing emissions by improving aircraft manufacturers' and operators' and policy makers' ability to assess the environmental benefits and costs of alternative policy measures. Such an assessment can then lead to the selection of the alternative that will achieve the greatest net environmental benefits. For example, one technology might greatly increase fuel efficiency, but produce higher nitrogen oxide emissions than another, somewhat less fuel-efficient technology. Overall, a cost benefit analysis might indicate that the less fuel-efficient technology would produce greater net benefits for the environment.

FAA currently supports several recent federal efforts to better quantify aviation emissions and their impact through improvements in emissions measurement techniques and modeling capability. One of these efforts is FAA's Partnership for Air Transportation and Emissions Reduction (PARTNER) Center of Excellence.[21] Created in 2003, PARTNER carries on what representatives of airlines, aircraft and engine manufacturers, and experts in aviation environmental research have described as a robust research portfolio. This portfolio includes efforts to measure aircraft emissions and to assess the human health and welfare risks of aviation emissions and noise. For example, researchers are developing an integrated suite of three analytical tools—the Environmental Design Space, the Aviation Environmental Design Tool, and the Aviation Environmental Portfolio Management Tool – that can be used to identify interrelationships between noise and emissions. Data from these three tools, together with the Aviation Environmental Design tool being developed by the Volpe National Transportation Systems Center and others, will allow for assessing the benefits and costs of aviation environmental policy options. Another R&D initiative, the Airport Cooperative Research Program (ACRP),[22] conducts applied research on aviation emissions and other environmental issues facing airports. The program is managed by the National Academies of Science through its Transportation Research Board under a contract with FAA, which provided $10 million for the program in both 2007 and 2008 and is seeking to increase these investments through its reauthorization to specifically focus on aviation environmental issues. Several of the emissions-related projects undertaken through ACRP have concentrated on developing methods to measure particulate matter and hazardous air pollutants at airports in order to identify the sources of these pollutants and determine whether their levels could have adverse health effects. FAA has also developed an Aviation Emissions Characterization roadmap to provide a systematic process to enhance understanding of aviation's air quality emissions, most notably particulate matter and hazardous air pollutants. In addition, FAA, in conjunction with NASA and the National Oceanic and Atmospheric Administration, launched the Aviation Climate Change Research Initiative to develop the scientific understanding necessary for informing efforts to limit or reduce aviation greenhouse gas emissions.

Another effort, the Commercial Aviation Alternative Fuels Initiative (CAAFI),[23] led by FAA, together with airlines, airports, and manufacturers, is intended to identify and eventually develop alternative fuels for aviation that could lower emissions of greenhouse gases, and other pollutants; increase fuel efficiency; and reduce U.S. dependence on foreign oil. CAAFI supports research on low-carbon fuel from sources such as plant oils, algae, and biomass that are as safe as petroleum-based fuel and compare favorably in terms of environmental impact. Part of the research will involve assessing the environmental impact of alternative fuels to determine whether their use could reduce emissions of pollutants that affect climate and air quality. The research will also assess the impact of producing these fuels on the overall carbon footprint. The CAAFI sponsors have set goals for certifying a 50 percent synthetic fuel for aviation use in 2008, a 100 percent synthetic fuel for use by 2010, and a biofuel made from renewable resources such as palm, soy, or algae oils. As part of CAAFI, Virgin Atlantic Airlines, together with Boeing, has tested a blend of kerosene (normal jet fuel) and biofuels in a flight from London to Amsterdam, and Continental, in association with Boeing and jet engine manufacturer General Electric, is planning a similar test in 2009.

NASA Conducts Fundamental Aeronautics R&D in Support of NextGen, Including Efforts That Can Help Lower Emissions

NASA has devoted a substantial portion of its aeronautical R&D program to the development of technologies critical to the implementation of NextGen, as well as new aircraft and engine technologies, both of which can help reduce aviation emissions.

NASA has three main aeronautics research programs – Fundamental Aeronautics, Aviation Safety, and Airspace Systems – each of which contributes directly and substantially to NextGen. For example, the Airspace Systems program supports research on air traffic management technologies for NextGen, and the Fundamental Aeronautics program focuses on removing environmental and performance barriers, such as noise and emissions, that could constrain the capacity enhancements needed to accommodate projected air traffic increases. Appendix II describes in more detail how NASA's aeronautics R&D programs support the implementation of NextGen.

NASA also works with aircraft and aircraft engine manufacturers to increase fuel efficiency and reduce emissions. Their efforts have contributed to a number of advancements in aircraft engine and airframe technology, and NASA's R&D on emissions-reduction technologies continues. NASA has set technology-level goals for reducing greenhouse gases, nitrogen oxides, and noise, which have become part of the U.S. National Aeronautics Plan. For example, the plan includes a goal for developing technologies that could reduce nitrogen oxide emissions during landings and takeoffs by 70 percent[24] below the ICAO current standard. The plan also sets a goal of increasing fuel efficiency (and thereby decreasing greenhouse gases emissions) by 33 percent. These technologies would be incorporated in the next generation of aircraft, which NASA refers to as N+1,[25] by 2015. However, as NASA officials note, these goals must be viewed within the context that each of the goals can be fully met only if it is the only goal. For example, the goal for reducing nitrogen oxides can be fully achieved only at the expense of the goals for lowering greenhouse gas emissions and noise, because it is technologically challenging to design aircraft that can simultaneously reduce all of these environmental impacts.

For the longer term (2020), NASA is focusing on developing tools and technologies for use in the design of advanced hybrid-wing body aircraft, the following generation of aircraft, or N+2. Emissions from these aircraft would be in the range of 80 percent below the ICAO standard for nitrogen oxide emissions during landings and takeoffs, and fuel consumption would be 40 percent less than for current aircraft. The U.S. aircraft and engine manufacturing industry has also set goals for reducing aircraft emissions in the engines the industry plans to produce. According to the Aerospace Industries Association, which represents this industry, its members have set a goal of reducing carbon dioxide emissions by 15 percent in the next generation of aircraft while continuing to significantly reduce nitrogen oxide emissions and noise.

The development of aircraft technologies such as those that NASA is currently working on to reduce emissions can take a long time, and it may be years before the technologies are ready to be incorporated into new aircraft designs. According to FAA, the development process generally takes 12 to 20 years. For example, the latest Pratt and Whitney engine, the geared turbofan, which is expected to achieve significant emissions and noise reductions, took 20 years to develop.

Several Steps Can Be Taken to Help Reduce Aviation Emissions, but Challenges Remain to Be Addressed

Reducing aviation emissions includes steps that FAA and others can take to move the implementation of NextGen forward and support R&D on NextGen and emissions-reduction technologies, as well as technical, financial, regulatory challenges facing the federal government, the aviation industry, and Congress.

Expediting the Implementation of NextGen Can Help Reduce Aviation Emissions

Implementing NextGen expeditiously is essential to handle the projected growth in air traffic efficiently and safely, and in so doing, help to reduce aircraft emissions. Steps to advance NextGen's implementation include management improvements and the deployment of available NextGen components.

Management improvements can move nextgen forward more efficiently

Several management actions are important to advance the implementation of NextGen. One such action is to establish a governance structure within FAA that will move NextGen initiatives forward efficiently and effectively. FAA has begun to establish a governance structure for NextGen, but it may not be designed to give NextGen initiatives sufficient priority to ensure the system's full implementation by 2025. Specifically, FAA's implementation plan for NextGen is called the Operational Evolution Partnership (OEP). The manager responsible for OEP is one of nine Vice Presidents who report to the Chief Operating Officer (COO) of FAA's Air Traffic Organization (ATO), who reports directly to the FAA Administrator. While the manager responsible for OEP is primarily responsible for implementing NextGen, other Vice Presidents are responsible for NextGen-related activities in their designated areas. In addition, the FAA managers responsible for airports and aviation safety issues are Associate Administrators who report through the Deputy FAA Administrator to the FAA Administrator. Some of the activities for which these Associate Administrators are responsible are critical to NextGen's implementation, yet there is no direct line of authority between the OEP manager and these activities.

Some congressional leaders and other stakeholders, including aviation industry representatives and aviation experts, view FAA's management structure for NextGen as too diffuse. Some of the stakeholders have called for the establishment of a position or NextGen program office that reports directly to the FAA Administrator to ensure accountability for NextGen results. These stakeholders have expressed frustration that a program as large and important as NextGen does not follow the industry practice of having one person with the authority to make key decisions. They point out that although the COO is nominally in charge of NextGen, the COO must also manage FAA's day-to-day air traffic operations and may therefore not be able to devote enough time and attention to managing NextGen. In addition, these stakeholders note that many of NextGen's capabilities span FAA operational units whose heads are on the same organizational level as the head of OEP or are outside ATO, and they believe that an office above OEP and these operational units is needed. In prior work, we have found that programs can be implemented most efficiently when managers are empowered to make critical decisions and are held accountable for results.[26]

Another management action is needed to help ensure that FAA acquires the skills required for implementation, such as contract management and systems integration skills. Because of the scope and complexity of the NextGen implementation effort, FAA may not have the in-house expertise to manage it without assistance. In November 2006, we recommended that FAA examine its strengths and weaknesses and determine whether it has the technical expertise and contract management expertise that will be needed to define, implement, and integrate the numerous complex programs inherent in the transition to NextGen.[27] In response to our recommendation, FAA has contracted with the National Academy of Public Administration (NAPA) to determine the mix of skills and number of skilled persons, such as technical personnel and program managers, needed to implement the new OEP and to compare those requirements with FAA's current staff resources. In December 2007, NAPA provided FAA with its report on the types of skills FAA will require to implement NextGen, and it has undertaken a second part of the study that focuses on identifying any skill gaps between FAA's current staff and the staff that would be required to implement NextGen.[28] NAPA officials told us that they expect to publish the findings of the second part of the study in the summer of 2008. We believe this is a reasonable approach that should help FAA begin to address this challenge as soon as possible. It may take considerable time to select, hire, train, and integrate into the NextGen initiative what could be a large number of staff.

We have also identified potential approaches for supplementing FAA's capabilities, such as having FAA contract with a lead systems integrator (LSI)–that is, a prime contractor who would help to ensure that the discrete systems used in NextGen will operate together and whose responsibilities may include designing system solutions, developing requirements, and selecting major system and subsystem contractors.[29] However, this approach would require careful oversight to ensure that the government's interests are protected and could pose significant project management and oversight challenges for the Joint Planning and Development Office (JPDO), the organization within FAA responsible for planning NextGen, and for FAA.

Deploying available NextGen components can demonstrate their ability to operate together and achieve anticipated efficiencies

Moving from planning to implementing some components of NextGen can begin to demonstrate the potential of the system as well as reduce congestion in some areas of the country, thereby also reducing emissions. Many of the technologies and procedures planned for NextGen are already available, and a few have been implemented individually, such as the CDA procedures in use in Los Angeles and Louisville and ADS-B in Alaska. However, the available technologies and procedures have not yet been deployed simultaneously to demonstrate that they can be operated safely as an integrated suite of technologies and procedures in the national airspace system. Several stakeholders have suggested that FAA consider a gradual rollout of NextGen technologies and procedures in a particular area. For example ADS-B technologies, CDA and RNAV/RNP procedures, and high-density airport operations could be deployed in a defined area of the current system, possibly in sequence over time, to test their combined use and demonstrate the safety of an integrated suite of NextGen advancements. Such a graduated rollout is sometimes referred to as "NextGen Lite." FAA is currently considering a demonstration project in Florida and Georgia, in which it, together with aviation equipment manufacturers and municipalities, would use the NextGen capabilities of ADS-B, RNAV, and RNP for on-demand air taxi fleet[30] operations. As other NextGen capabilities, such as System-Wide Information Management (SWIM),[31] are deployed and as air taxi fleet operations move to other airports and regions, the demonstration will be expanded to include those new capabilities and other airports and regions. According to the airlines and other stakeholders we interviewed, a demonstration of the successful integration of NextGen capabilities and of efficiencies resulting from their use would give the airlines an incentive to equip their aircraft with NextGen technologies. They could then lower their costs by reducing their fuel consumption and decrease the impact of their operations on the environment. The findings from our research indicate that such regional or targeted demonstrations could accelerate the delivery of NextGen benefits while helping to ensure safe operations within the current system. In addition, demonstrations can increase stakeholders' confidence in the overall NextGen initiative.

Table 1. The President's Budget for NASA's Aeronautics Programs for Fiscal Years 2007 and 2008 and Budget Projections for Fiscal Years 2009 through 2013

(Dollars in millions)

	Fiscal year						
	Enacted		Requested	Proposed			
Program	**2007**	**2008**	**2009**	**2010**	**2011**	**20112**	**2013**
Aviation Safety							
Integrated Vehicle Health Management	30.7	22.2	19.7	19.9	18.8	18.6	19.2
Aging Aircraft	14.9	10.0	10.6	11.3	11.2	12.0	12.4
Integrated Resilient Aircraft Control	22.2	15.3	17.1	18.5	19.0	18.2	18.8
Integrated Intelligent Flight Deck Technologies	19.5	19.3	15.2	16.3	16.0	15.7	16.1
Subtotal	**87.3**	**66.5**	**62.6**	**65.9**	**65.0**	**64.5**	**66.5**
Airspace Systems							
NextGen – Airspace	85.1	83.3	61.3	56.0	57.3	58.5	60.8
NextGen – Airportal	17.4	16.8	13.3	16.7	16.9	16.9	17.5
Subtotal	**102.5**	**100.1**	**74.6**	**72.7**	**74.2**	**75.4**	**78.4**
Fundamental Aeronautics							
Subsonic –Rotary Wing	36.1	30.8	25.8	26.6	26.7	26.9	28.0
Subsonic – Fixed Wing	133.9	119.9	108.4	105.3	107.6	109.1	111.5
Supersonics	67.7	53.0	44.0	44.9	44.3	45.2	46.6
Hypersonics	92.8	66.2	57.3	56.4	56.5	57.4	58.4
Subtotal	**330.4**	**269.9**	**235.4**	**233.2**	**235.2**	**238.6**	**244.6**
Aeronautics Test Program							
Aero Ground Test Facilities	48.5	50.0	48.2	49.4	50.8	51.0	51.0
Flight Operations and Test Infrastructure	25.0	25.1	25.6	26.4	27.2	27.2	27.2
Subtotal	73.5	75.1	73.9	75.8	78.0	78.2	78.2
Total	**593.8**	**511.7**	**446.5**	**447.5**	**452.4**	**456.7**	**467.7**

Source: NASA.

Note: Most of the research on aircraft emissions reductions that NASA performs is funded through the Fundamental Aeronautics – Fixed Wing program.

Resolving Aeronautics R&D Funding Issues Is a Further Step in Addressing Aviation Emissions

Federal funding for aeronautics research, the category that includes work on aviation emissions, has declined over the past decade, particularly for

NASA, which historically provided most of the funding for this type of research. NASA's current aeronautics research budget is about half of what it was in the mid-1990s. Moreover, the budget request for aeronautics R&D for fiscal year 2009 is $447 million, or about 25 percent less than the $594 million provided in fiscal year 2007. (See table 1.) According to NASA, about $280 million of the proposed $447 million would contribute to NextGen. In addition, according to NASA officials, a significant portion of the funding for subsonic fixed-wing aircraft is directed toward emissions-related research, and many other research efforts contribute directly or indirectly to potential emissions-reduction technologies.

As its funding for aeronautics R&D has declined, NASA has emphasized fundamental research, which serves as the basis for developing technologies and tools that can later be integrated into aviation systems, and has focused less on developmental and demonstration work. As a result, NASA is now sometimes developing technologies to a lower maturity level than in the past, and the technologies are less ready for manufacturers to adopt them, resulting in a gap in the research needed to bring technologies to a level where they can be transferred to industry for further development. Failure to address this gap could postpone the development of emissions-reduction technologies.

As a partial response to the gap, the administration has proposed some additional funding for FAA that could be used to further develop NASA's and others' emissions- and noise reduction technologies. Specifically, FAA's reauthorization proposal seeks $111 million through fiscal year 2011 for the CLEEN Engine and Airframe Technology Partnership,[32] which FAA officials said is intended to provide for earlier maturation of emissions and noise technologies while NASA focuses on longer-term fundamental research on noise and emissions. The CLEEN partnership, which is also contained in the House's FAA reauthorization bill,[33] would create a program for the development and maturation of certifiable engine and airframe technologies for aircraft over the next 10 years which would reduce aviation noise and emissions. The legislation would require the FAA Administrator, in coordination with the NASA Administrator, to establish objectives for developing aircraft technology outlined in the legislation. The technology requested to be developed would increase aircraft fuel efficiency enough to reduce greenhouse gas emissions by 25 percent relative to 1997 subsonic jet aircraft technology, and, without increasing other gaseous or particle emissions, reduce takeoff-cycle nitrogen oxide emissions by 50 percent relative to ICAO's standard. Although FAA's reauthorization bill has not yet been enacted, the administration's proposed fiscal year 2009 budget includes \$10 million for the CLEEN program.

The CLEEN program would be a first step toward further maturing emissions and noise reduction technologies, but experts agree that the proposed funding is insufficient to achieve needed emissions reductions. While acknowledging that CLEEN would help bridge the gap between NASA's R&D and manufacturers' eventual incorporation of technologies into aircraft designs, aeronautics industry representatives and experts we consulted said that the program's funding levels may not be sufficient to attain the goals specified in the proposal. According to these experts, the proposed funding levels would allow for the further development of one or possibly two projects. Moreover, in one expert's view, the funding for these projects may be sufficient only to develop the technology to the level that achieves an emissions-reduction goal in testing, not to the level required for the technology to be incorporated into a new engine design. Nevertheless, according to FAA and some experts we consulted, the CLEEN program amounts to a pilot project, and if it results in the development of emissions-reduction technologies that can be introduced into aircraft in the near future, it could lead to additional funding from the government or industry for such efforts.

FAA and NASA have identified the R&D that is needed for NextGen, but have not determined what needs to be done first, at what cost, to demonstrate and integrate NextGen technologies into the national airspace system. Completing this prioritization is critical to avoid spending limited funds on lower-priority efforts or conducting work out of sequence. Once the identified R&D has been prioritized and scheduled, cost estimates can be developed and funds budgeted. Prioritizing research needs is an essential step in identifying the resources required to undertake the research.

The European Union is investing substantially in R&D that can lead to fuel-efficient, environmentally friendly aircraft. In February 2008, the European Union announced the launch of the Clean Sky Joint Technology Initiative, with total funding of \$2.4 billion over 7 years—the European Union's largest-ever research program. The initiative establishes a Europe-wide partnership between industry, universities, and research centers and aims to reduce aircraft emissions of carbon dioxide and nitrogen oxides by up to 40 percent and aircraft noise levels by 20 decibels. According to FAA, it is difficult to compare funding levels for U.S. and European R&D efforts because of differences in program structures and funding mechanisms, Nevertheless, foreign government investments of such magnitude in R &D on environmentally beneficial technologies could reduce the competitiveness of the U.S. aircraft manufacturing industry, since greater investments are likely to lead to greater improvements in fuel efficiency and keep U.S. aircraft manufacturers competitive in the global economy as well as reducing aviation's impact on the environment.

Reducing the Impact of Aviation Emissions Poses Technical, Financial, and Regulatory Challenges

Reducing aviation emissions will require technological advances, the integration of lower-emitting aircraft and NextGen technologies into airline fleets, and strengthened or possibly new regulations to improve air quality and limit greenhouse gas emissions. Fulfilling these requirements will pose challenges to aviation because of the technical difficulties involved in developing technologies that can simultaneously address air pollutants, greenhouse gases, and noise; constraints on the airline industry's resources to invest in new aircraft and technologies needed to reduce emissions and remain competitive; and the impact that emissions regulations can have on the aviation system's expansion and the financial health of the aviation industry.

Simultaneously Addressing Air Pollutants, Greenhouse Gases, and Noise from Aircraft Presents Technical Challenges

Although the aviation industry has made strides in lowering emissions, more reductions are needed to keep pace with the projected growth in aviation, and achieving these reductions will be technically challenging. NASA's efforts to improve jet engine designs illustrate this challenge: While new designs have increased fuel efficiency, reduced most emissions, and lowered noise, they have not achieved comparable reductions in nitrogen oxide emissions. Nitrogen oxide emissions have increased because new aircraft engines operate at higher temperatures, producing more power with less fuel and lower carbon dioxide and carbon monoxide emissions, but also producing higher nitrogen oxide emissions, particularly during landings and takeoffs, when engine power settings are at their highest. It is during the landing/takeoff cycle that nitrogen oxide emissions also have the greatest impact on air quality. As discussed, nitrogen oxides contribute to ground-level ozone formation. Similarly, as we noted in a report on NASA's and FAA's aviation noise research earlier this year,[34] it is technologically challenging to design aircraft engines that simultaneously produce less noise and fewer greenhouse gas and other emissions. Although it is possible to design such engines, the reductions in greenhouse gases could be limited in engines that produce substantially less noise. NASA and industry are working on technologies to address these environmental trade-offs. For example, the Pratt & Whitney geared turbo fan engine that we mentioned earlier is expected to cut nitrogen oxide emissions in half while also improving fuel efficiency and thereby lowering carbon dioxide emissions. Nevertheless, it remains technologically challenging to design aircraft that can reduce one environmental concern without increasing another.

In a 2004 report to Congress on aviation and the environment,[35] FAA noted that the interdependencies between various policy, technological, and operational options for addressing the environmental impacts of aviation and the full economic consequences of these options had not been appropriately assessed. However, in recent years, FAA has made progress in this area, including its sponsorship of the previously mentioned PARTNER study on the interrelationships between noise and emissions. This study can be used to assess the costs and benefits of aviation environmental policy options.

The Financial Condition of the Airline Industry Creates a Challenge to Implementing Emissions-Reduction Technologies

Most U.S. airlines have stated that they plan to invest in aircraft and technologies that can increase fuel efficiency and lower emissions, but in the near term, integrating new aircraft into the fleet, or retrofitting aircraft with technologies that can improve their operational efficiency, poses financial challenges to the airline industry. Aircraft have an average lifespan of about 30 years, and the airlines can take almost that entire period to pay for an aircraft. The current fleet is, on average, about half as many years old—11 years for wide-body aircraft, and 14 years for narrow- body aircraft—and therefore is expected to be in operation for many years to come. In addition, the financial pressures facing many airlines make it difficult for them to upgrade their fleets with new, state-of-the-art aircraft, such as the Boeing 787 and Airbus A380, which are quieter and more fuel efficient, emitting lower levels of greenhouse gases.[36] Currently, U.S. carriers have placed a small proportion (40, or less than 6 percent) of the over 700 orders that Boeing officials say the company has received for its 787 model. Furthermore, no U.S. carriers have placed orders for the new Airbus 380. These financial pressures also limit the airlines' ability to equip new and existing aircraft with NextGen technologies such as ADS-B that can enable more efficient approaches and descents, resulting in lower emissions levels. FAA estimates that it will cost the industry about $14 billion to equip aircraft to take full advantage of NextGen.

Delays by airlines in introducing more fuel-efficient, lower-emitting aircraft into the U.S. fleet and in equipping or retrofitting the fleet with the technologies necessary to operate NextGen could limit FAA's ability to efficiently manage the forecasted growth in air traffic. Without significant reductions in emissions and noise around the nation's airports, efforts to expand their capacity could be stalled and the implementation of NextGen delayed because of concerns about the impact of aviation emissions. As we previously reported,[37] offering operational advantages, such as preferred takeoff and landing slots, to fuel-efficient, lower-emitting aircraft or aircraft equipped with ADS-B could create incentives for the airlines to invest in the necessary technologies. Similarly, as noted, deploying an integrated suite of NextGen technologies and procedures in a particular region could create incentives for carriers to equip their aircraft with NextGen technologies.

More Stringent Regulatory Standards Pose Challenges for Airport Expansion Projects

Concerns about the health effects of air pollutants have led to more stringent air quality standards that could increase the costs or delay the implementation of airport expansion projects. In recent years, EPA has been implementing a more stringent standard for ozone emissions to better protect the health of people exposed to it, and this standard could require more airports to tighten controls on nitrogen oxides and some types of volatile organic compounds that also contribute to ozone formation. Under the current standard,[38] 122 airports are located in areas that are designated as nonattainment areas. This number includes 43 of the 50 busiest U.S. commercial service airports. In March 2008, EPA further revised the ozone standard, because new evidence demonstrated that exposure to ozone at levels below the level of the current standard are associated with a broad array of adverse health effects.[39] This recent revision to the ozone standard will increase the number of U.S. counties, and hence airports, that will be in nonattainment. EPA estimated that the number of affected counties could potentially grow from 104 to 345 nationwide. While the exact number of airports that will be affected has not been officially determined at this time, FAA estimates that a modest number of commercial service airports in California, Arizona, Utah, Texas, Oklahoma, Arkansas, and along the gulf coast to Florida will be in nonattainment areas for the revised 8-hour ozone standard. According to EPA, any development project beginning in 2011 at these airports would have to conform to the state implementation plan.

As communities gain more awareness of the health and environmental effects of aviation emissions, opposition to airport expansion projects, which has thus far focused primarily on aviation noise, could broaden to include emissions. According to a California air quality official, many of the same communities that have interacted with airports over aviation noise have more recently recognized that they could also be affected by emissions from airport sources. In Europe, concerns about the impact of aviation on air quality and climate change have led to public demands for tighter control over aircraft emissions, and these demands have hindered efforts to expand airports in Birmingham, and London (Heathrow). Moreover, a plan to expand London's Stansted Airport was rejected because of concerns about climate change that could result from additional emissions.

To minimize constraints on the future expansion of airport capacity stemming from concerns about the health and environmental effects of aviation emissions, it will be important for airports; the federal and state governments; and the airline industry to work together to accurately characterize and address these concerns and to take early action to mitigate emissions. As noted, constraints on efforts to expand airports or aviation operations could affect the future of aviation because the national airspace system cannot expand as planned without a significant increase in airport capacity. The doubling or tripling of air traffic that FAA expects in the coming decades cannot occur without additional airports and runways.

Market-Based Initiatives to Reduce Aviation Emissions of Greenhouse Gases Could Pose Challenges for U.S. Airlines by Increasing Their Costs

Concerns about the environmental effects of greenhouse gas emissions have grown steadily over the years, leading to national and international efforts to limit them. In the

In the United States, EPA has not regulated greenhouse gas emissions;[40] however, Congress is taking steps to deal with climate change, some of which could include market-based measures that would affect the aviation industry. For example, several bills were introduced in the 110th Congress to initiate cap and trade[41] programs for greenhouse gas emissions[42] None of these bills would include aviation directly in a cap and trade program. However, some could have indirect consequences for the aviation industry by, for example, requiring fuel producers to purchase allowances through the system to cover the greenhouse gas content of the fuel they sell to the aviation sector. The cost of purchasing these allowances could be passed on to fuel consumers, including airlines, raising the cost of jet fuel. Fuel is already the airline industry's largest cost. According to the Air Transport Association, cap and trade programs that significantly increase airline fuel costs could have significant consequences for the industry and such programs could make it more difficult for carriers to pay for aircraft or technologies that would reduce greenhouse gas emissions. As we have previously noted,[43] cap and trade programs can cost-effectively reduce emissions of greenhouse gases such as carbon dioxide, especially when compared with other regulatory programs. However, it is important that the impact of such measures on various sectors of the economy, such as the aviation industry, be thoroughly considered.

Internationally, ICAO has not set standards for aircraft carbon dioxide emissions,[44] but it has been working, with the support of FAA, other government aviation authorities, and the aviation industry, to develop a strategy for addressing the impact of aviation on climate change, among several efforts to address climate change. For example, ICAO published a manual for countries, Operational Opportunities to Minimize Fuel Use and Reduce Emissions. In 2004, ICAO endorsed the development of an open emissions trading system as one option countries might use and endorsed draft guidance for member states on establishing the structural and legal basis for aviation's participation in a voluntary open trading system. The guidance includes information on key elements of a trading system, such as reporting, monitoring, and compliance, while encouraging flexibility to the maximum extent possible. In adopting the guidance last fall at the ICAO Assembly, all 190 Contracting States with the exception of those in the European Union—agreed that the inclusion of one country's airlines in another country's emissions trading system should be based on mutual consent between governments.

Consistent with the requirement to pursue reductions of greenhouse gas emissions from international aviation through ICAO, some countries that have included the aviation sector in their emissions trading systems or other emissions-reduction efforts have, excluded international flights. Consequently, these countries' efforts will not affect U.S. airlines that fly into their airports. The European Union (EU), however, is developing legislation, which has not been finalized, that would include both domestic and international aviation in an emissions trading scheme.[45] As proposed, the EU's scheme would apply to air carriers flying within the EU and to carriers, including U.S. carriers, flying into and out of EU airports in 2012. For example, under the EU proposal, a

U.S. airline's emissions in domestic airspace as well as over the high seas would require permits if a flight landed or departed from an EU airport. Airlines whose aircraft emit carbon dioxide at levels exceeding prescribed allowances would be required to reduce their emissions or to purchase additional allowances. Although the legislation seeks to include U.S. airlines within the emissions trading scheme, FAA and industry stakeholders have argued that U.S. carriers would not legally be subject to the legislation.

While the EU's proposal to include international aviation in its emissions trading system is intended to help forestall the potential catastrophic effects of climate change, according to FAA and airlines, it will also affect the aviation industry's financial health. In particular, according to FAA and airline and aircraft and engine manufacturing industry representatives, the EU's proposal could disadvantage U.S. airlines, which have older, less fuel- efficient fleets than their European competitors. Paying for emissions credits could, according to U.S. airlines, also leave them with less money for other purposes, including investing in newer, more fuel-efficient aircraft and technologies to improve flight efficiency and reduce fuel usage. Furthermore, according to U.S. carriers, the proposed trading scheme unfairly penalizes the aviation sector because it lacks a readily available non-carbon-based alternative fuel, whereas other sectors can use alternative fuels to reduce their emissions.

The governments of many nations, including the United States, oppose the European Union's proposal to unilaterally include international aviation in its emissions trading system because the proposed approach is not consistent with ICAO guidance. Furthermore, such an approach could be inconsistent with international aviation agreements and may not be enforceable. According to FAA, the EU's inclusion of aviation in its emissions trading scheme violates the Chicago Convention on International Civil Aviation[46] and other international agreements. FAA further notes that the EU proposal ignores differences in the U.S. and EU aviation systems[47] and ignores a performance-based approach in which countries decide which measures are most appropriate for goals on emissions. We are currently undertaking for this Subcommittee a study of the EU emissions trading system and its potential impact on U.S. airlines, and other issues relating to aviation and climate change.[48]

Mr. Chairman, this concludes my prepared statement. I would be pleased to respond to any questions that you or other Members of the Subcommittee may have.

APPENDIX I. FEDERAL AGENCY VIEWS ON HEALTH AND ENVIRONMENTAL EFFECTS OF AIR POLLUTION

Pollutant	Heath effects	Environmental effects
Ozone	Lung function impairment, effectson exercise performance, increased airway responsiveness, increased susceptibility to respiratory infection, increased hospital admissionsand emergency room visits, pulmonary inflammation, and lung structure damage (long term).	Resultsfrom animal studiesindicate that repeated exposure to high levelsof ozone for several monthsor more can produce permanent structural damage in the lungs. Ozone isalso responsible for several billion dollarsof agricultural crop yield lossin the United Stateseach year.
Carbon monoxide	Most seriousfor those who suffer from cardiovascular disease. Healthy individualsare also affected, but only at higher levelsof exposure. Exposure to elevated carbon monoxide levelsisassociated with visual impairment, reduced work capacity, reduced manual dexterity, poor learning ability, and difficulty in performing complex tasks.	Adverse health effectson animalssimilar to effectson humans.
Nitrogen oxides	Lung irritation and lower resistance to respiratory infections.	Acid rain, visibility degradation, particle formation. Contribute stoward ozone formation, and acts as a greenhouse gas in the atmosphere and, therefore, may contribute to climate change.
Particulate matter	Effectson breathing and respiratory systems, damage to lung tissue, cancer, and premature death. The elderly, children, and people with chronic lung disease, influenza, or asthma, tend to be especially sensitive to the effectsof particulate matter.	Visibility degradation, damage to monumentsand buildings, safety concernsfor aircraft from reduced visibility.
Volatile organic compounds	Eye and respiratory tract irritation, headaches, dizziness, visual disorders, and memory impairment.	Contribute to ozone formation, odors, and have some damaging effect on buildingsand plants.
Carbon dioxide, water vapor, and contrails	None.	Act asgreenhouse gasesin the atmosphere and, therefore, may contribute to climate change. Contrailsand contrail-induced cloudsproduce warming effect regionally where aircraft fly.
Sulfur dioxide	Breathing, respiratory illness, alterationsin pulmonary defenses, and aggravation of existing cardiovascular disease.	Together, sulfur dioxide and nitrogen oxidesare the major pre cursors to acid rain, which isassociated with the acidification of lakes and streams, accelerated corrosion of buildingsand monuments, and reduced visibility.

Sources: EPA and FAA.

Appendix II. Examples of the National Aeronautics and Space Administration's Research and Development Programs Supporting NextGen

NextGen research and development (R&D) needs	NextGen capabilities from the National Aeronautics and Space administration's (NASA) R&D programs
Safety management procedures that can predict, rather than respond to, safety risks, in ahigh density, complex operating environment; research to support safety analysis, development of advanced materials for continued air worthiness of aircraft, aircraft system and equipage management; and adaptive aircraft control systemsto allow the crew and aircraft to recover from unsafe conditions.	Under itsAviation Safety program, NASA research supports development of Safety Management Systemsto provide asystematic approach to manage safety risks; integrates prediction and mitigation of risks prior to aircraft accidents or incidents; and sharessafety-related information through programs such asthe Aviation Safety Analysis and Information Sharing program.
Improved air traffic management technologies to manage airspace configuration, support increasesin volume and complexity of traffic demands, mitigate weather impacts, and maintain safe and efficient operations at airports, decrease runway incursions, and address wake vortex issues.	Under its Airspace Systems program, NASA research supports development of variable separation standards based on aircraft performance levelsin the en route environment; trajectory-based operations, traffic spacing, merging, metering, flexible terminal airspace, and expanded airport access; technologies and procedures for safe runway proceduresin low-visibility conditions;coordinated arrival/departure management; and mitigation of weather and wake vortex issues.
Management of aviation growth to meet the complexity of operations within the Next Gen environment, regulation and certification of new manned and unmanned aircraft, and management of operationsin an environmentally sound manner.	Under itsFundamental Aero nautics program, NASA research supports development of improved performance for the next generation of conventional subsonic aircraft, rotorcraft and supersonic aircraft and develops methods for environmental management system to measure and assessreductionsin air quality impact, noise, and emissions.

Source: GAO analysis of Joint Planning and Development Office and NASA information.

End Notes

[1] These emissions include airborne pollutants, which affect air quality, and greenhouse gases, primarily carbon dioxide, which are produced by the combustion of fossil fuel, and contribute to climate change.

[2] These figures are based on a long-range FAA forecast using 2006 as the baseline.

[3] See the list of related products at the end of this statement, especially GAO, *Next Generation Air Transportation System: Progress and Challenges in Planning and Implementing the Transformation of the National Airspace System,* GAO-07-649T (Washington, D.C.: Mar. 22, 2007.)

[4] As we noted in our recent testimony before this Subcommittee, aviation noise has been a greater constraint on airport expansion efforts than aviation emissions, but we are limiting our discussion in this testimony to aviation emissions.

[5] 42 U.S.C. §4332(2)(C).

[6] States are required to submit implementation plans to EPA for reducing emissions in areas that fail to meet the National Ambient Air Quality Standards set by EPA under the Clean Air Act for common air pollutants with health and environmental effects (known as criteria pollutants). Geographic areas that have levels of a criteria pollutant above those allowed by the standard are called nonattainment areas.

[7] 42 U.S.C. §7506(c)(1) (The Conformity Provision).

[8] ICAO is an organization affiliated with the United Nations that aims to promote the establishment of international civilian aviation standards and recommended practices and procedures. FAA, as the U.S. representative to ICAO, in consultation with EPA, works with representatives from other countries to formulate aircraft emissions standards.

[9] See the list of related GAO products at the end of this statement, especially GAO, *Aviation and the Environment: Strategic Framework Needed to Address Challenges Posed by Aircraft Emissions,* GAO-03-252 (Washington, D.C.; Feb. 28, 2003).

[10] We are currently undertaking a study on aviation environmental trends, efforts, and challenges for this Subcommittee and the Subcommittee on Space and Aeronautics, Committee on Science and Technology, House of Representatives.

[11] Motor vehicles include cars and buses for airport operations and passenger, employee, and rental agency vehicles.

[12] Ground service equipment includes aircraft tugs, baggage and belt loaders, generators, lawn mowers, snow plows, loaders, tractors, air-conditioning units, and cargo moving equipment.

[13] Hazardous air pollutants from aviation activities include benzene and formaldehyde.

[14] Ground-level ozone is formed when nitrogen oxides and volatile organic compounds as well as other gases and substances are mixed and heated in the atmosphere.

[15] Contrails are clouds and condensation trails that form when water vapor condenses and freezes around small particles (aerosols) in aircraft exhaust.

[16] Intergovernmental Panel on Climate Change, *Aviation and the Global Atmosphere* (1999).

[17] Intergovernmental Panel on Climate Change, Fourth Assessment Report, *Summary for Policy Makers,* Cambridge University Press, Cambridge, UK, November 2007.

[18] A critical component of RNP is the ability of the navigation system to monitor its achieved navigation performance and to identify for the pilot if an operational requirement is or is not being met during an operation.

[19] GAO-03-252.

[20] Alternatively, some scientists studying options for addressing climate change believe that a price on emissions would represent the most effective means of achieving reductions overall.

[21] FAA Centers of Excellence are FAA partnerships with universities and affiliated industry associations and businesses throughout the country that conduct aviation research in a number of areas, including advanced materials, aircraft noise, and aircraft emissions.

PARTNER is a cooperative research organization that includes 10 collaborating universities and approximately 50 advisory board members who represent aerospace manufacturers, airlines, airports, state and local governments, and professional and community groups. NASA, FAA, and Transport Canada are sponsors of PARTNER. The collaborating universities and organizations represented on the advisory board provide equal matches for federal funds for research and other activities.

[22] ACRP was authorized in 2003 as part of Vision 100—Century of Aviation Reauthorization Act, Pub. L. 108-176, Section 712 (Dec 12, 2003).

[23] CAAFI, established in October 2006, is sponsored by the Air Transport Association, the Aerospace Industries Association, and the Airports Council International-North America under the direction of FAA, and involves stakeholders from industry, universities, and other federal agencies, including NASA.

[24] This goal is at a pressure ratio of 30, over the ICAO standard adopted at the Committee on Aviation Environmental Protection's sixth meeting (CAEP 6), with commensurate reductions over the full pressure ratio range.

[25] "N" refers to the current generation of tube-and-wing aircraft entering service in 2008, such as the Boeing 787.

[26] See GAO, *Best Practices: Better Support of Weapon System Program Managers Needed to Improve Outcomes*, GAO-06-110 (Washington, D.C.: Nov. 30, 2005). In this study of private-sector best practices that could be applied to federal programs,[26] we found that program managers at highly successful companies were empowered to decide whether programs were ready to move forward and to resolve problems and implement solutions. In addition, program managers were held accountable for their choices.

[27] GAO, *Next Generation Air Transportation System: Progress and Challenges Associated with the Transformation of the National Airspace System,* GAO-07-25 (Washington, D.C.: Nov. 13, 2006).

[28] NAPA, *Workforce Needs Analysis for the Next Generation Air Transportation System (NEXTGEN): Preliminary Findings and Observations* (Washington, D.C.: December 2007).

[29] GAO-07-25.

[30] Air taxis are small aircraft that can be hired to carry passengers or cargo and are regulated under Part 135 of the Federal Aviation Regulations.

[31] SWIM is information-management architecture for the national airspace system, acting as its "World-Wide Web." SWIM will manage surveillance, weather, and flight data, as well as aeronautical and system status information and will provide the information securely to users.

[32] CLEEN stands for continuous lower energy emissions and noise.

[33] H.R. 2881.

[34] GAO, *Aviation and the Environment: Impact of Aviation Noise on Communities Presents Challenges for Airport Operations and Future Growth of the National Airspace System,* GAO-08-216T (Washington, D.C.: Oct. 24, 2007).

[35] FAA, *Aviation and the Environment: A National Vision Statement, Framework for Goals and Recommended Actions* (Washington, D.C.: December 2004).

[36] We are currently undertaking a study for this Subcommittee and the House Committee on Transportation and Infrastructure that, among other things, will assess the financial condition of the airlines.

[37] GAO, *Aviation and the Environment: FAA'S and NASA's Research and Development Plan's for Noise Reduction Are Aligned, but the Prospects of Achieving Noise Reduction Goals Are Uncertain,* GAO-08-384 (Washington, D.C.: Feb. 15, 2008).

[38] In 2003, EPA began implementing a new standard that called for concentrations of ozone not to exceed 0.08 parts per million over an 8-hour period. The former standard required concentrations not to exceed 0.12 parts per million over a 1-hour period. The more stringent standard resulted in the designation of more nonattainment areas for ozone. These areas contained 12 airports.

[39] 73 Fed. Reg. 16436 (Mar. 27, 2008). The new standard would lower the allowed concentrations of ozone from 0.08 parts per million in an 8-hour period to 0.075 parts per million during that period.

[40] Recently, however, the Supreme Court ruled that greenhouse gases meet the Clean Air Act's definition of an air pollutant and that EPA has the statutory authority to regulate greenhouse gas emissions from new motor vehicles under the Clean Air Act. *Massachusetts v. Environmental Protection Agency*, 127 S.Ct. 1438, 1459-62 (2008). As a result of this opinion, EPA must take one of three actions: (1) issue a finding that greenhouse gas emissions cause or contribute to air pollution that may endanger public heath or welfare; (2) issue a finding that greenhouse gases do not endanger public health or welfare; or (3) provide a reasonable explanation as to why it cannot or will not exercise its discretion to issue a finding. If EPA makes an endangerment finding, the Clean Air Act requires EPA to regulate greenhouse gas emissions from new motor vehicles. In response to this case, EPA has announced that it will issue an Advance Notice of Proposed Rulemaking on "specific effects of climate change and potential regulation of greenhouse gas emissions from stationary and mobile sources under the Clean Air Act.

[41] Cap and trade programs combine a regulatory limit or cap on the amount of a substance—in this case a greenhouse gas such as carbon dioxide—that can be emitted into the atmosphere with market elements like credit trading to give industries flexibility in meeting this cap. A current example is the cap and trade program for sulfur dioxide under the Clean Air Act. This program includes electric utilities, which are the primary emitters of sulfur dioxide, and established a cap on the utilities' emissions. Sulfur dioxide allowances were primarily given (rather than auctioned) to companies.

[42] S. 28, S, 309, S. 317, S. 485, S. 1168, S. 1177, S. 1201, S. 1554, S. 1766, S. 2191,H.R. 620, H.R. 1590, H.R. 3989, H.R. 4226.

[43] GAO, *Vehicle fuel Economy: Reforming Fuel Economy Standards Could Help Reduce Oil Consumption by Cars and Light Trucks, and Other Options Could Complement These Standards,* GAO-07-921 (Washington, D.C.: Aug. 2, 2007).

[44] According to FAA, the last extensive discussion within ICAO on carbon dioxide emissions from aircraft occurred several years ago. At that time, ICAO's experts agreed that the cost of fuel provided sufficient incentive to minimize fuel consumption – hence carbon dioxide emissions. There was some technical work around 2001 on the development of an aircraft efficiency parameter, which might have been used to target carbon dioxide reductions. However, it failed to identify a parameter that would be able to assess aircraft fleets in their multiple operational environments in an equitable manner.

[45] The emissions trading scheme involves a cap and trade system that sets allowances for greenhouse gas emission for industries and other sources. Parties that pollute below their allowance receive emissions credits, which they can trade in a market to other parties that have exceeded their allowance.

[46] The Chicago Convention on International Civil Aviation of 1944 organized global aviation. According to the Convention, no state may condition the right of transit over or entry into or exit from its territory of any aircraft of another state on their operator's payment of fees, dues, or other charges.

[47] For example, FAA notes that there are considerable differences in the air traffic system efficiencies across the Atlantic, that the United States has a domestic fuel tax while nearly all EU states have none, and that the cost of fuel is about 50 percent more expensive for U.S. airlines because of the dollar's weakness in recent years.

[48] This ongoing work was jointly requested by the Committee on Transportation and Infrastructure, House of Representatives, and the Committee on Science and Technology, House of Representatives.

In: Aviation and Climate Change
Editors: George T. Blumenthal
ISBN: 978-1-60876-757-1

Chapter 9

AVIATION AND CLIMATE CHANGE

James E. McCarthy

SUMMARY

Aircraft are a significant source of greenhouse gases—compounds that trap the sun's heat, with effects on the Earth's climate. In the United States, aircraft of all kinds are estimated to emit between 2.4% and 3.4% of the nation's total greenhouse gas (GHG) emissions, depending on whether one counts international air travel. The impact of U.S. aviation on climate change is perhaps twice that size when other factors are considered. These include the contribution of aircraft emissions to ozone formation, the water vapor and soot that aircraft emit, and the high altitude location of the bulk of aircraft emissions. Worldwide, aviation is projected to be among the faster growing GHG sources.

If Congress and the new Administration decide to regulate aircraft GHG emissions, they face several choices. The Administration could use existing authority under Section 231 of the Clean Air Act, administered by the Environmental Protection Agency. EPA has already been petitioned to do so by several states, local governments, and environmental organizations. Congress could address aviation or aviation fuels legislatively, through cap-and-trade or carbon tax proposals.

Among the legislative options, the cap-and-trade approach (setting an economy-wide limit on GHG emissions and distributing tradable allowances to emitters) has received the most attention. In the 110th Congress, most cap-and-trade bills would have included aviation indirectly, through emission caps imposed upstream on their source of fuel—the petroleum refining sector. By capping emissions upstream of air carriers and eventually lowering the cap as much as 80%, bills such as these would have several effects: they would provide an incentive for refiners to produce lower carbon fuels; they would increase the price of fuels, and thus increase the demand for more fuel-efficient aircraft; and they might increase the cost of aviation services relative to other means of transport, giving airline passengers and shippers of freight incentives to substitute lower-cost, lower carbon alternatives.

Besides regulating emissions directly or through a cap-and-trade program or carbon tax, there are other tools available to policymakers that can lower aviation's GHG emissions. These include implementation of the Next Generation Air Traffic Control System (not expected to be complete until 2025, although some elements that could reduce aircraft emissions may be implemented sooner); research and development of more fuel-efficient aircraft and engines; and perhaps the development of lower carbon jet fuel.

This report provides background on aviation emissions and the factors affecting them; it discusses the tools available to control emissions, including existing authority under the Clean Air Act and proposed economy-wide cap-and-trade legislation; and it examines international regulatory developments that may affect U.S. commercial airlines. These include the European Union's Emissions Trading Scheme for greenhouse gases (EU-ETS), which is to include the aviation sector beginning in 2012, and discussions under the auspices of the International Civil Aviation Organization (ICAO).

INTRODUCTION

Research on climate change has identified a wide array of sources that emit "greenhouse gases" (GHGs)—compounds that trap the sun's heat, with effects on Earth's climate.[1] The largest sources of these emissions, particularly in developed economies, are electric utilities and the transportation sector.[2] In the United States, electricity generation accounts for about 40% of the emissions of carbon dioxide, the principal greenhouse gas, or about one-third of the emissions of the six major GHGs combined.[3] The transportation sector, including cars, trucks, buses, trains, ships, and aircraft, accounts for roughly one-third of U.S. CO_2 emissions, or 27% of the six GHGs combined.

AIRCRAFT EMISSIONS

Aircraft account for about 10% of the U.S. transportation sector's GHG emissions, or 2.4% to 3.4% of total U.S. GHG emissions, but they have been among the faster growing sources of GHG emissions worldwide. According to the Commission of the European Union, emissions from international aviation increased by almost 70% between 1990 and 2002.[4] The United Nations Intergovernmental Panel on Climate Change (IPCC), in a 1999 study that is still widely cited, projected that the impact of aircraft emissions on climate would be 2.6 to 11 times as large in 2050 as it was in 1992.[5] If, as many argue, GHG emissions must be reduced 50% to 80% in that time period, emissions from aviation would need to be drastically reduced to provide a proportional share of the targeted reduction.

U.S. emissions from aircraft have run counter to the worldwide trends and projections, however. Since 1990, aircraft GHG emissions have declined both as a percentage of total U.S. emissions and in absolute terms (see Table 1). The biggest factor in the decline was a 56 percent decrease in emissions from domestic military operations, which more than offset a small increase in domestic commercial and general aviation[6] emissions.

Table 1. CO_2 Emissions from U.S. Aviation, 1990-2006 (million metric tons of CO_2 equivalent)

Fuel / Aircraft Type	1990	2000	2006
Jet Fuel			
Commercial Aircraft	136.7	164.2	142.1
Military Aircraft	33.9	20.5	14.8
General Aviation	6.3	9.2	11.4
Aviation Gasoline			
General Aviation	3.1	2.5	2.3
Total	180.0	196.4	170.6
% of Total U.S. GHG Emissions	2.9%	2.8%	2.4%

Source: U.S. EPA, *Inventory of U.S. Greenhouse Gas Emissions and Sinks: 1990-2006*

Emissions from domestic operation of commercial aircraft grew 4% between 1990 and 2006. That figure was well below the growth in air travel: according to the Air Transport Association (the association that represents the domestic airlines) passenger-miles traveled domestically on U.S. commercial airlines increased 74% between 1990 and 2006.[7]

Two types of efficiency increases contributed to the slow growth in U.S. commercial aircraft emissions. First, load factors (the percentage of seats occupied) increased to 79.9% in 2007, compared with 60.4% in 1990. Second, fuel efficiency itself increased, as older, less efficient aircraft were retired in favor of newer, more efficient models. These savings can be substantial. For example, American Airlines estimates that the 18-year old MD-80s currently flying use 35% more fuel than the Boeing 73 7-800 aircraft that are to replace them over the next two years.[8]

EPA's *Inventory of U.S. Greenhouse Gas Emissions and Sinks* shows that domestic flights of all kinds (military, commercial aircraft, and general aviation) accounted for about 10% of the GHG emissions from the U.S. transportation sector in 2006—2.4% of overall U.S. GHG emissions. Aviation's impact on climate may be greater than these figures suggest, however, for two reasons. First, emissions resulting from international transportation are not currently included in the U.S. emission totals.[9] These emissions grew from 45.7 million metric tons in 1990 to 71.1 million in 2006, a 56% increase. If they were included in the U.S. aviation statistics, U.S. emissions from aircraft of all types would have risen 7% since 1990 (rather than falling 5%), and aviation would account for 3.4% of the U.S. GHG total. Second, the bulk of the aviation sector's emissions occur high in the atmosphere, where their impact on climate is greater than that of emissions at ground level. According to a number of sources, the total impact of aviation could be around twice the impact of carbon dioxide alone when this factor is taken into account.[10] Emissions from jet aircraft also lead to the formation of cirrus clouds, as the condensation trails (contrails) of water vapor and sulfur particles emitted from engines at high altitudes form ice crystals that persist as clouds under some atmospheric conditions. Scientists are uncertain how to measure the occurrence and impact of such clouds, but they are reasonably certain that the clouds add to the greenhouse effect of aircraft emissions, perhaps substantially.[11]

Thus, while the precise share of aviation in total greenhouse gas emissions depends on what is included, and the impact of some emissions is unclear, there is little doubt that aviation is a significant contributor to U.S. and world GHG emissions.[12]

Reducing Emissions: Non-Regulatory Factors

Fuel Cost

The cost of jet fuel represents a significant portion of total cost for most air carriers. There is a great deal of variation depending on the distance traveled, the age and efficiency of the aircraft, and the price of fuel at any given time, but the total fuel expenses of U.S. airlines consumed an average of 24% of airline operating revenues in 2007, according to the Air Transport Association. This percentage is expected to be even higher in 2008: although costs are now declining rapidly, in the first nine months of the year, domestic airlines spent $46 billion on fuel for their domestic and international flights, 60% more than the cost in 2007.[13]

Given the importance of fuel costs, airlines and air freight companies have a major incentive to purchase more fuel-efficient aircraft, and thus, aircraft manufacturers are constantly seeking to improve the efficiency of airplanes and engines. These incentives have resulted in sizeable efficiency gains: U.S. airlines carried 20.4% more passenger and cargo traffic in 2007 than they did in 2000, but they used nearly 3% less fuel in doing so. This resulted in a reduction of 5.1 million metric tons of CO_2 emissions in 2007, as compared to 2000, according to ATA.[14] The industry has committed to a further 30% increase in fuel efficiency by 2025.[15]

Air Traffic Control

In addition to improving the efficiency of individual aircraft, there is a general consensus that fuel use could be reduced by modernizing the Federal Aviation Administration (FAA)'s air traffic control system. The FAA is in the process of transforming air traffic control from a ground-based system of radars to a satellite-based system, dubbed the Next Generation (NextGen) Air Transportation System. The primary objective is to enable the air traffic control system to handle a projected doubling of current passenger loads by 2025. But, when fully implemented, NextGen is also expected to cut the GHG emissions of individual aircraft 10% to 15%, by allowing more direct routing, delay reduction, and through such features as Continuous Descent Approach.[16] According to the FAA, United Parcel Service aircraft equipped with some of the NextGen technologies have reduced emissions as much as 34%.[17]

REGULATING AIRCRAFT UNDER THE CLEAN AIR ACT

As policy-makers consider whether the federal government should regulate aircraft GHG emissions (versus continuing to rely solely on market forces to determine the level of emissions), some have turned their attention to the potential for regulation under the Clean Air Act. In December 2007, EPA received two petitions requesting that it exercise that authority to regulate GHG emissions from aircraft engines.[18]

To date, EPA has not promulgated regulations to control CO_2 from *any* source. In 2003, responding to an earlier petition to regulate GHGs from cars and trucks, the agency maintained that it did not have authority under the Clean Air Act to do so. That determination was challenged by Massachusetts and other petitioners, and in a 2007 decision, the U.S. Supreme Court found that GHGs *are* air pollutants within the Clean Air Act's definition, and thus, EPA would have authority to regulate them if it so chose.[19]

Under Section 231 of the Act, the EPA Administrator may propose emission standards applicable to any air pollutant from any class of aircraft engines which in the Administrator's judgment causes, or contributes to, air pollution which may reasonably be anticipated to endanger public health or welfare. The Administrator is required to consult with the FAA Administrator and hold public hearings before finalizing such standards. The President may disapprove of such standards if the Secretary of Transportation finds that they would create a hazard to aircraft safety.

The December 2007 petitions request that EPA make a finding that aircraft GHG emissions do endanger public health or welfare, and that the agency adopt regulations that allow a range of compliance approaches: these might include emission limits, operational practices, fees, a capand-trade system, minimizing engine idling time, employing single engine taxiing, or use of ground-side electricity measures to replace the use of fuel-burning auxiliary power units at airport gates.[20]

The decisions about whether and how to regulate aircraft might seem relatively straightforward, but the new Administration actually faces broader decisions than whether it should regulate GHGs from aircraft. For one thing, the aircraft petitions are among several others that EPA has received to regulate GHG emissions from power plants, cars and trucks, ships, and other sources. As a result, whatever decision is made (for any one of these sectors) is likely to affect the decisions regarding all the others—ultimately, a large portion of the economy. Given the relative size of aircraft emissions as compared to, say, power plants or cars and trucks, aviation would seem an unlikely choice to be the first sector whose GHG emissions EPA would regulate.

In addition, although the President-elect has said that he favors controls on GHG emissions, among the questions he faces will be whether to proceed with such regulation under the existing Clean Air Act authority or to support legislation targeted more specifically at GHGs. New legislation might be more efficient—clearly allowing sources in different industries to trade emission allowances to each other, for example—and it might avoid challenge in the courts if Congress were specific regarding the authority it was giving EPA to control GHG emissions. The current language of the act, while arguably providing regulatory authority, is sufficiently vague that legal challenges are considered almost a certainty if EPA proceeds. This might delay implementation of controls.

The two options—proceeding under the Clean Air Act or supporting new legislation—are not mutually exclusive. Existing EPA authority under the Clean Air Act could be used as a backstop, while Congress considers granting new authority. If EPA moves ahead with regulations, it could motivate Congress and interested parties to agree on a legislative approach. By holding open the possibility that regulation of some sort is a certainty, the development of regulations at EPA might speed the enactment of new, perhaps more flexible and more cost-effective, legislative authority.

PROPOSED CAP-AND-TRADE LEGISLATION

Whether or not the new Administration moves forward with GHG regulations, GHG legislation is considered a high priority of the new Congress, with attention centered on legislation that would cap emissions of GHGs economy-wide and establish an allowance trading system for major emitters. (For a general discussion of how such cap-and-trade systems work, see CRS Report RL34502, *Emission Allowance Allocation in a Cap-and-Trade Program: Options and Considerations,* by Jonathan L. Ramseur, especially Appendix A.)

As noted, aviation is considered a significant source of GHG emissions. Despite this, the aviation sector (like other transportation sources) has not been targeted directly by the climate change bills introduced in Congress to date. In the 110th Congress, for example, the Lieberman-Warner bill (S. 2191/S. 3036), which was reported by the Senate Environment and Public Works Committee and debated briefly on the Senate floor before failing to achieve cloture, would have dealt with aviation emissions by including the refining sector in its overall emissions cap, thus regulating the aviation sector's emissions "upstream."[21] The same would have been the case under the draft legislation prepared by Representatives Boucher and Dingell, posted on the House Energy and Commerce Committee website in October, 2008, but not actually introduced.

By capping emissions from fuels upstream of the air carriers and eventually lowering the cap as much as 80%, bills such as these would have several effects: first, they would provide an incentive for refiners to produce lower carbon fuels[22]; second, they would increase the price of fuels, as refiners either purchased additional allowances for their emissions or were forced to reduce production, in essence rationing fuels through a higher price in order to stay beneath the emissions cap; third, as the cost of fuel increased, the demand for fuel-efficient aircraft would increase; and fourth, consumers of aviation services (airline passengers and shippers of freight) would have incentives to replace higher-cost air transportation with lower-cost alternatives (e.g., video-conferencing by business and government entities, increased reliance on lower-emission forms of transport, greater reliance on local sources of goods, etc.).

The cost of air travel and of air freight has been reduced substantially since its inception, as aircraft have become more efficient and airlines have reduced other costs in competitive markets. According to ATA, the cost of domestic air travel in real (inflation-adjusted) terms has declined by 51.9% since 1978.[23] By contrast, controlling GHG emissions, if it were done, would likely increase the price of air travel and air freight, reducing demand in comparison to a business as usual (i.e., without GHG controls) scenario.

International Developments

European Union

Unlike the upstream approach of U.S. cap-and-trade bills, the European Union (EU) has chosen to regulate aviation directly, by including the sector in its Emission Trading Scheme (ETS), beginning in 2012. The ETS began operation in 2005, capping emissions of CO_2 from more than 10,000 energy-intensive stationary sources of emissions. The currently covered sectors (power plants; petroleum refining; iron and steel production; coke ovens; pulp and paper; and cement, glass, lime, brick, and ceramics production) account for about half of EU CO_2 emissions.[24]

On January 1, 2012, the aviation sector's CO_2 emissions are to be added to the ETS. The scheme is to cover all aircraft operators landing in or departing from the EU, with the exception of military aircraft, some small carriers, emergency services, research, and humanitarian flights. Thus, flights to and from the EU by U.S. air carriers would be subject to the emission limits. For the first year, the total quantity of allowances would be equivalent to 97% of the sector's average 2004-2006 emissions. The cap would be reduced to 95% in 2013, with further reductions to be agreed on as part of the ongoing review of the ETS. In allocating the emissions allowed under the cap, 85% of the sector's 2012 allowances are to be given to aircraft operators at no cost, and 15% of the allowances auctioned. The EU Commission has proposed that 80% of allowances be distributed free of charge in 2013, with 20% being auctioned; the percentage of free allowances is expected to continue declining with a goal of auctioning all allowances in 2020.

Operators emitting more than their allowed cap would need to buy additional allowances on the carbon market. A special reserve fund, taken from the sector's overall cap, is to allocate up to one million tons worth of allowances a year to ensure access to the market to new operators and to provide allowances to rapidly growing airlines.[25]

The directive provides sanctions for failure to comply with the scheme, including the possibility that a non-complying airline might be banned from operating in the EU.[26] According to press reports, "This warning shot is aimed at foreign airlines—including US carriers—that have said they will not recognise the scheme."[27] For its part, the United States has responded by threatening trade sanctions if the EU makes an attempt to force foreign airlines to comply with the emissions trading system.[28]

This dispute highlights a general problem in directly regulating emissions from sectors such as aviation or shipping, a substantial portion of whose total emissions occur outside of national borders. Without international agreement, it may be difficult to enforce emission limits, and the imposition of controls by any one country or bloc of countries is likely to be challenged through existing international institutions.

Regulating upstream of the aviation industry, as the climate change bills of the 110th Congress would have done, may avoid some of these issues, maintaining a level playing field for U.S. and foreign airlines and air freight companies, without imposing emission limits that could be directly challenged or circumvented. Whether enactment of such legislation would be sufficient to address European concerns over U.S. airlines' emissions, resolving the dispute, remains to be seen.

ICAO

The EU is not the only international body addressing aircraft emissions. The International Civil Aviation Organization (ICAO), the international organization that administers standards and recommended practices for the aviation authorities of more than 190 countries, agreed in September 2007 to support the development of an "aggressive" action plan on aviation and climate change, but without a fixed timetable or specific emission reduction targets.[29] The United States has supported the ICAO as the proper venue for international regulation of emissions, and maintains that the EU's approach is contrary to ICAO's charter, the Chicago Convention on International Civil Aviation.[30] A majority of ICAO's members agree with the United States that participation in an emissions trading scheme (such as EU-ETS) should only be on the basis of mutual consent.[31]

CONCLUSION

Greenhouse gas emission controls of some sort may affect U.S. aviation in the next few years, be they specific controls on engine emissions, emission caps applied to the sector as a whole, upstream caps (on fuel refiners), or carbon taxes.[32] Depending on their stringency, the effects of most of these approaches could ripple through the economy, providing additional incentives for aircraft manufacturers to improve the fuel efficiency of aircraft, raising the cost of air travel and air freight, and providing further pressure to improve the air traffic control system.

U.S. airlines and air freight companies, like many other sectors, would prefer that they be allowed to address the GHG issue through voluntary measures. Unlike some other sectors, they have achieved substantial increases in fuel economy over the last three decades or more, and have stabilized their GHG emissions at roughly their 1990 levels. But the sector is still an important source of emissions, and its projected growth indicates that it may outstrip the economy as a whole's rate of emission growth, particularly if one includes emissions from international flights; thus, it is likely to be included in some fashion in any mandatory economy-wide approach to reducing GHG emissions.

On a practical level, reducing emissions from aviation may be more complicated than it is for most other sectors:

- The sector is composed of tens of thousands of mobile emission sources; thus, direct controls on engines or aircraft face obstacles that do not apply in industries composed of fewer and stationary emission sources. Even monitoring the relevant emissions for this sector is difficult.
- The sector's emissions affect climate in several ways. Controlling only CO_2 emissions might leave other impacts of aircraft on climate unaffected. More research is needed to identify the precise effects of some of these, such as the impact of contrails on cirrus cloud formation, and the effect of such clouds on climate change.
- The sector's impressive progress in making itself more energy-efficient in recent years poses obstacles as well: improving load factors was relatively easy when they were at 60%; at the current level, 80%, one begins to approach the limits of further improvement.
- Some means of emission reduction are beyond the industry's control, including the pace of modernization of the air traffic control system, and the degree to which aeronautical research and engine modifications can reduce fuel consumption. In both cases, emission reduction may depend, at least in part, on the actions of government agencies—the FAA and NASA, in particular. According to ATA, funding for NASA and FAA aviation environmental R&D programs has been cut by approximately 50 percent in the past 10 years.[33]
- Finally, the sector faces controls from foreign countries, particularly the European Union. International negotiations for a post-Kyoto-Protocol emissions control scheme may give rise to emission limits in other countries, as well.

As discussed, Congress and the new Administration have a number of options, including several forms of legislation; regulation by EPA under the existing Clean Air Act is another possibility. If the Administration so chooses, the existing Clean Air Act might prove a particularly important tool to bring interested parties to the table, while providing a backdrop to consideration of legislation by Congress.

End Notes

[1] Six greenhouse gases are the primary focus of concern: carbon dioxide (CO_2), methane (CH_4), nitrous oxide (N_2O), sulfur hexafluoride (SF_6), hydrofluorocarbons, and perfluorocarbons. These six are the subject of international agreements (the U.N. Framework Convention on Climate Change and its Kyoto Protocol) and are the emissions that would be subject to control in most climate change cap-and-trade bills that have been introduced in Congress. As will be noted later in this report, other emissions from aircraft, especially water vapor and the persistent condensation trails (contrails) that form in jet engine exhaust, may have an impact on climate as well, but in general they have not been the subject of negotiations, international agreements, or legislation.

[2] For data on these and other sectors, see "Trends in Greenhouse Gas Emissions," Chapter 2 of U.S. EPA, *Inventory of U.S. Greenhouse Gas Emissions and Sinks: 1990-2006*, at *http://www.epa.gov/climatechange/emissions/* usinventoryreport.html, especially Table 2-1.

[3] Because each gas has a different heat-trapping potential (e.g., methane has 25 times the heat-trapping potential of CO_2, and SF_6—although emitted in small quantities—has 22,800 times CO_2's heat trapping potential), GHG emissions are generally converted to tons of CO_2 equivalent in order to assess the climate change contribution that an economic sector makes.

[4] See Europa, website of the European Commission's Directorate General for Environment, "Aviation and Climate Change" http://ec.europa.eu/environment. Visited November 12, 2008.

[5] IPCC, *Aviation and the Global Atmosphere*, Summary for Policy Makers, 1999, at http://www.ipcc.ch/ipccreports/ sres/aviation/008.htm. The term for its impact is "radiative forcing."

[6] The term "general aviation" refers to flights other than those by the military, scheduled commercial airlines, and large air cargo operators.

[7] Data on load factors and revenue passenger miles (as well as other industry data) are available from the Air Transport Association's *2008 Economic Report*, at http://www.airlines.org /NR/rdonlyres/770B5715-5C6F-44AA-AA8CDC9AEB4E7E12/0/2008AnnualReport.pdf for the period 1997-2007. Information for 1990 were provided in personal communications from ATA staff.

[8] "American Speeds Plans to Phase Out Old Planes," *Greenwire*, August 14, 2008.

[9] The UN Framework Convention on Climate Change refers to such emissions as combustion of "international bunker fuel," and categorizes the emissions separately from national emission totals, pending further agreement on how to address related emissions.

[10] See, for example, Testimony of Dr. David W. Fahey, Office of Oceanic and Atmospheric Research, National Oceanic and Atmospheric Administration, at the Subcommittee on Aviation, House Committee on Transportation and Infrastructure hearing on Aviation and the Environment: Emissions, May 6, 2008. Dr. Fahey was a lead author of portions of the 1999 and 2007 IPCC reports that considered the impact of global aviation on climate.

[11] Ibid.

[12] Another source, a report prepared for the International Civil Aviation Organization (ICAO) in 1999, said, "Aircraft are estimated to contribute about 3.5 per cent of the total radiative forcing (a measure of change in climate) by all human activities and ... this percentage, which excludes the effects of possible changes in cirrus clouds, was projected to grow." ICAO, "Environmental (ENV) Unit, Aircraft Engine Emissions, Definition of the Problem," at http://www.icao.int/cgi/goto_m_atb.pl?/icao/en/env/aee.htm, visited November 6, 2008. Similar conclusions were reached by the Federal Aviation Administration, which estimates that emissions of CO_2 and NOx from domestic aircraft will increase 60% by 2025. See FAA, *Aviation and Emissions: A Primer*, January 2005, p. 10, http://www.faa.gov/ regulations

[13] ATA, "Monthly Jet Fuel Cost and Consumption Report," at *http://www.airlines.org/economics* MonthlyJetFuel.htm.

[14] ATA, *2008 Economic Report*, previously cited, p. 25.

[15] Ibid., p. 17.

[16] Continuous Descent Approach, in which an aircraft lands by descending at a constant 3-degree angle rather than descending and holding at a series of altitude "steps," lowers fuel use and emissions by shortening flight time and eliminating the need for engine thrust required in a stepped approach to landing.

[17] See FAA, "Fact Sheet: Next Generation Air Transportation System 2006 Progress Report," October 10, 2007, at http://www.faa.gov/news/fact_sheets/news_story.cfm?news Id=8336.

[18] The first petition, submitted December 4, 2007, was filed by California, Connecticut, New Jersey, New Mexico, the Pennsylvania Department of Environmental Protection, New York City, the District of Columbia, and California's South Coast Air Quality Management District (the air pollution control agency for the Los Angeles area). The second petition was filed December 31, 2007, by Earthjustice on behalf of four environmental organizations.

[19] Massachusetts v. EPA, 127 S. Ct. 1438 (2007).

[20] For a brief discussion of the petitions, see 73 *Federal Register* 44460, July 30, 2008. Some of these measures, such as minimizing engine idling time, employing single engine taxiing, and use of ground-side electricity measures to replace the use of fuel-burning auxiliary power units, are already widely used by the airlines as fuel-saving measures.

[21] The Lieberman-Warner bill also provided for a National Academy of Sciences study of the aviation sector's emissions, including the identification of existing best practices to reduce emissions, recommendations for research for technologies and operations with the highest potential to reduce emissions, and recommendations of actions that the Federal Government could take to encourage or require additional emission reductions.

[22] Besides incentives, S. 2191/S. 3036 actually would have required the production of low carbon fuels, although it is not clear whether the requirement would have applied to aviation fuels. For further discussion, see CRS Report RL34489, *Climate Change: Costs and Benefits of S. 2191/S. 3036*, by Larry Parker and Brent D. Yacobucci, pp. 54-55. The Boucher-Dingell draft would have given EPA authority to limit lifecycle GHG emissions from transportation fuels, although, as with the Lieberman-Warner bill, it is not clear that that authority would have applied to aviation fuels.

[23] ATA, *2008 Economic Report*, previously cited, p. 11.

[24] For a description of the EU ETS, see CRS Report RL34150, *Climate Change and the EU Emissions Trading Scheme (ETS): Kyoto and Beyond*, by Larry Parker.

[25] "Airline Emissions Covered in EUETS from 2012," *ENDS Report*, July 2008, p. 51. The text of the ETS amendment adding the aviation sector to the scheme can be found at http://www.europarl.europa.eu/sides/getDoc.do?pubRef=- //EP//TEXT+TA+P6-TA-2008-0333+0+DOC+XML. The agreement on an EU-wide directive does not complete the legislative process. The 27 Member States now have until July 2009 to "transpose" the directive into national laws.

[26] The threat of a ban is found in Whereas clause (26): "In the event that an aircraft operator fails to comply with the requirements of this Directive and other enforcement measures by the administering Member State have failed to ensure compliance, Member States should act in solidarity. The administering Member State should therefore be able to request the Commission to decide on the imposition of an operating ban at Community level on the aircraft operator concerned, as a last resort."

[27] "Airline Emissions Covered in EUETS from 2012," *ENDS Report*, July 2008, p. 51.

[28] "Aviation and Emissions Trading," July 10, 2008, EurActive.com at *http://www.euractiv. com/en/climate-change/* aviation-emissions-trading/article-139728.

[29] "ICAO Backs Mutual Agreement Approach to Emissions Reductions but Europe Objects," *Daily Environment Report*, October 1, 2007.

[30] "Emissions Trading: EU Lawmakers Back Plan to Add Aviation To Emissions Trading Scheme in 2012," *Daily Environment Report*, July 9, 2008.

[31] ICAO, *Annual Report of the Council, 2007*, pp. 41-42.

[32] Carbon taxes are not discussed in this report, but their effects might be similar to the imposition of an upstream cap on emissions. They would raise the cost of fuel, thus encouraging the development of more fuel-efficient and lower carbon alternatives.

[33] ATA, *2008 Economic Report*, previously cited, p. 17. The FAA's efforts on NextGen have already been discussed. For an overview of aeronautics research goals, in which NASA plays a leading role, see National Science and Technology Council, *National Plan for Aeronautics Research and Development and Related Infrastructure*, December 2007,especially pp. 50-52, at *http://www.aeronautics. nasa.gov/ releases/aero_rd_ plan_ final_21_dec_2007.pdf.*

CHAPTER SOURCES

The following chapters have been previously published:

Chapter 1 – This is an edited, excerpted and augmented edition of a United States Government Accountability Office (GAO), Report to Congressional Committees. Publication GAO-09-554, dated June 2009.

Chapter 2 – This is an edited, excerpted and augmented edition of a Federal Aviation Administration, Office of Environment and Energy publication, dated January 2005.

Chapter 3 - These remarks were delivered as Statement of Richard L. Altman, Exective Director, Commercial Aviation Alternative Fuels Initiative (CAAFI), before the House Transportation Committee, U.S. House of Representatives, dated May 6, 2008.

Chapter 4 – These remarks were delivered as Statement of James K. Coyne, President, National Air Transportation Association, before the Subcommittee in Aviation, Committee on Transportation and Infrastructure, U.S. House of Representatives, dated May 6, 2008.

Chapter 5 – These remarks were delivered as Statement of Daniel K. Elwell, Assisstant Administrator, Office of Aviation Policy, Planning and Environment, Federal Aviation Administration, before the House Transportation and Infrastructure Committee, Subcommittee on Aviation, U.S. House of Representatives, dated May 6, 2008.

Chapter 6 – These remarks were delivered as Written Statement of Dr. David W. Fahey, Earth System Research Laboratory, Office of Oceanic and Atmospheric Research, National Oceanic and Atmospheric Administration, U.S. Department of Commerce, before the House Transportation and

Infrastructure Committee, Subcommittee on Aviation, U.S. House of Representatives, dated May 6, 2008.

Chapter 7 - These remarks were delivered as Statement of James K. May, President and CEO, Air Transport Association of America, Inc., before the House Transportation and Infrastructure Committee, Subcommittee on Aviation, U.S. House of Representatives, dated May 6, 2008.

Chapter 8 – These remarks were delivered as Statement of Gerald L. Dillingham, Ph.D., Director, Physical Infrastructure Issues, before the House Transportation and Infrastructure Committee, Subcommittee on Aviation, U.S. House of Representatives, dated May 6, 2008.

Chapter 9 - This is an edited, excerpted and augmented edition of a United States Congressional Research Service publication, Report Order Code R40090, dated December 16, 2008.

INDEX

D

E

F

G

H

I

P

Q

R

S

T